The Close-Combat Files of Colonel Rex Applegate

Col. Rex Applegate & Maj. Chuck Melson

Paladin Press • Boulder, Colorado

The Close-Combat Files of Colonel Rex Applegate
by Col. Rex Applegate and Maj. Chuck Melson

ISBN 0-87364-998-2
Printed in the United States of America

Published by Paladin Press, a division of
Paladin Enterprises, Inc.
Gunbarrel Tech Center
7077 Winchester Circle
Boulder, Colorado 80301, USA.
+1.303.443.7250

Direct inquiries and/or orders to the above address.

Visit our Web site at www.paladin-press.com

CONTENTS

The personal weapons and equipment carried by OSS operative T/5 John W. Brunner in the China-Burma-India theater during World War II. (Photo courtesy of John W. Brunner.)

How Sleep the Brave

The only thing about this mission we
Disliked was knowledge that our failure to
Return would be a veil to hide our fate
Forever from the people who should know.
But even this could not put down our high
Excitement as we dropped into the pit
Of night and listened, after parachutes
Were open, to the fading motors of the plane,
Remembering the pilot's grin
And the good luck sign already a vision from the past.
Whatever ground it was that rose to meet
Our groping feet, you may be sure it gave
Us all the cloak and dagger stuff that we'd been itching for.
In secret files, in secret hearts.

—Joseph A. Bourdow
Infantry Journal, June 1948

ACKNOWLEDGMENTS

Additional information was obtained from David W. Arnold, Steve Barron, Clyde Beasley, Dr. John W. Brunner, Kathy Fotheringham, Chuck Karwan, Robert C. Kasper, David L. Kentner, Larry R. Kubacki, Lawrence H. McDonald, Dr. Gregory B. Morrison, William Pilkington, Peter R. Robins, Bruce K. Siddle, Nicholas S. Tyler, Maj. Edward J. Wages, the Catoctin Mountain National Park, the Federal Bureau of Investigation, the Fort Ritchie U.S. Army Garrison, the Franklin D. Roosevelt Library, the Imperial War Museum, the National Archives and Records Administration, and the U.S. Air Force Special Operations Command.

COAUTHOR'S NOTE

Documents and illustrations are from Rex Applegate's personal files unless otherwise indicated. Bibliographic information for published sources is provided in the initial entry with an abbreviated format used thereafter. Abbreviations used with notes and references include: ed., eds. (editor or editors); GHCA (Gung Ho Chuan Association); IWM (Imperial War Museum); NARA (National Archives and Records Administration); n.d. (no date); n.p. (depending on location, no place, no publisher, no page); ms (manuscript).

EDITOR'S NOTE

This unique book contains reproductions of many of Colonel Applegate's lesson plans, reports, memorandums, notes, and other writings. Very little editing has been done to them so that they would remain as much as possible in their original form.

WARNING

The information in this book is presented for historical purposes only and is not intended as a basis for instruction or action, particularly by unqualified individuals without proper facilities and expert supervision. Legislation governing actions people may take in defending themselves vary from state to state and county to county, and readers are advised to consult their particular jurisdiction's laws before attempting to use any of the material described herein. This book is *for academic study only*.

INTRODUCTION

COLONEL APPLEGATE ON SHOOTING FOR KEEPS

Point shooting with a handgun has been called by many names over the years. In World War II we called it "instinctive pointing." Others called it "point shoulder shooting." Some mistakenly called it "hip shooting." Recently, others have called it worse names than that, declaring that it was a hoax, and so on. Some of the so-called "gun gurus" claim that there has to be a "flash front sight" picture always present, even in the dark. Simply put, point shooting is the employment of the handgun in close-quarters combat without the need, or use, of the sights. So that you can better appreciate the importance of this type of shooting, this introduction will describe how it was developed. The actual technique of point shooting is covered within this text and elsewhere.

My earliest introduction to handgun shooting was probably fortunate, because I grew up influenced mainly by Gus Peret, a professional exhibition shooter for Remington-Peters, who was my uncle. He would come home to Yoncalla, Oregon, and do his practice shooting, much of which was of the exhibition type, such as that practiced by Annie Oakley, the Topperweins, Herb Parsons, and in the Buffalo Bill circus. These people did not use the sights for many of their acts. I threw the bricks in the air for Peret to shoot at while he practiced point, or instinctive, shooting. I also cleaned his guns and the range, so I was in hog heaven.

I did a lot of plinking and handgun hunting when younger, but I wasn't introduced to any formal type of target shooting or handgun training until I went to the University of Oregon and enrolled in the U.S. Army Reserve Officer Training Corps program. I entered active duty in 1940 at Ft. Lewis, Washington, as a second lieutenant, Military Police Corps, 3d Infantry Division.

By the time I had my first lieutenant bars, I had been given a regular U.S. Army commission. I had also become what the U.S. Army declared to be "thoroughly trained with the U.S. Army Model 1911 handgun." I vaguely thought, at the time, that shooting only at bull's-eyes was a pretty incomplete method of teaching a guy to train for battle. I might add that during that period everybody in law enforcement also trained by shooting, one-handed, at bull's-eye targets.

This became more evident when I was assigned to the Coordinator of Information (COI), which was the forerunner of the Office of Strategic Services (OSS). I was instructed by then Col. William J. "Wild Bill" Donovan to learn all there was to know about close combat with and without weapons.

I had access to pre–World War II books such as those by FitzGerald and McGivern. However, there were few written words on how the old frontier gunmen were supposed to have done it.

The first indication of how the old-timers did it was in a book I found called *Triggernometry*, written by a man named Eugene Cunningham. I went through this book several times and it gave me a few clues as to how they shot, but most was assumption. Remember, also, that this was the time of the old cowboy movie stars, such as William S. Hart, Tom Mix, and Hopalong Cassidy. These men depicted shooting from the hip, killing Indians and badmen by "snapping" the revolver and knocking them off of a galloping horse with a handgun, and so on. This kind of stuff was the public conception, my conception, and practically everyone else's of the way to shoot a handgun in close-quarters battle.

Because of *Triggernometry*, in 1942 I went to Deadwood, South Dakota, where James Butler "Wild Bill" Hickok was killed. When I visited Deadwood I found it was a very remote spot with none of the hype now associated with Hickok, who was buried there.

Wild Bill was an authentic Western gunman who actually killed a lot of men in combat; he wasn't a phony. By the time he was murdered, he was a national figure made famous by authors of dime novels like Ned Buntline. There were many stories about the mythical exploits of Hickok that had been vastly overinflated by the dime-novel press. However, he also had a documented record of killing a number of men in face-to-face gunfights in the Old West.

I was still searching for that essential fact: how did they do it? I got some answers when I went over to the old Deadwood County courthouse. There was a little old lady there and I said, "Ma'am, do you have any newspaper articles, clippings, or information on your most famous local character, Wild Bill Hickok?" She pondered a while and said, "Lieutenant, I think we have something down in the basement." She was gone 15 to 20 minutes and came back with a dusty bundle of papers with a big old red ribbon tied around it.

I had a few hours before the next train, so I started going through the papers. Most of them were just repeats of the dime-novel episodes or old newspaper stories. However, there were two letters fastened together by an old straight pin. One was a letter from an admirer writing to Hickok, asking, in effect, "How did you kill these men? What was your method or technique?" This was exactly what I was looking for. *What goes on in a gunfight?*

For some reason, Hickok's letter, in his own handwriting, was fastened to this letter of inquiry. Obviously, it had never been mailed. He wrote, "I raised my hand to eye level, like pointing a finger, and fired." This was contrary to all the Old West, Hollywood shoot-from-the-hip techniques.

This was very intriguing and interesting, but it wasn't made clear to me until I started my training in combat handgun shooting under a couple of gentlemen named William E. Fairbairn and Eric A. Sykes. For you people who may not know, they were the English police officers who came out of Shanghai during World War II. They were pioneers in many present-day police combat methods and Fairbairn was the father of the modern special weapons and tactics (SWAT) team.

While in the Far East, Sykes and Fairbairn also first developed successful police usage and training of the handgun in combat. They were coauthors of the first text on the subject, *Shooting to Live*, now reprinted and published by Paladin Press. They also designed the world famous Fairbairn-Sykes British commando knife.

These men operated in Shanghai from 1900 to 1940, when Shanghai was one of the most lawless areas of the world. Fairbairn rose from a constable to assistant commissioner of the Shanghai Municipal Police (SMP), its second highest rank. Not only did he have extensive command and training experience, but in his career he participated in and studied the results of more than 200 gun battles. He was a man who not only had experienced violence but was very curious about this aspect of police work and about why things worked, and why things did not. He developed many tactics and techniques over the early years of this century that are standard police practice today.

In 1940, shortly after World War II started, the British army left most of its armament and equipment on the French side of the English Channel at Dunkirk. The British were very ill-prepared and ill-equipped for the German *blitzkrieg* type of warfare, and the victorious Germans backed the British up to the beach.

The British government executed a rescue mission across the English Channel using all available boats, and most of the military personnel were rescued. Still, the British army and British people were without firearms. The problem was immense: Hitler was on the coast of France and plans were being made for an airborne invasion of England.

This was the time when Fairbairn and Sykes were called from Shanghai to offer their services and expertise based on the violence they had experienced in the Far East. Both were made captains in the British army the minute they stepped off the boat from Shanghai. Their first assignment was to train the British home guard and auxiliary units to defend against the perceived imminent German invasion. They had little to train with. The British people and police were almost entirely without weapons because of stringent gun control laws. The British army having left the bulk of its weaponry on the shores of France, the entire nation was absolutely helpless. Although the German invasion, expected at any time, did not materialize during 1940 and 1941, it could have. The two captains taught the home guard how to fight with scythes, pitchforks, and any improvised weapon, and how to conduct guerrilla operations in the countryside and streets.

Fairbairn and Sykes, although members of the British colonial police service, were considered pariahs by the British law enforcement establishment because they were so "into" violence. The English *bobby* was famous for, and bragged about, not carrying a firearm. The British attitude was *no crime in England; we don't need firearms, old boy*. British law enforcement generally considered them to be terrible people. (They were barbarians over there in Shanghai, *killing all those criminals*.) Fairbairn and Sykes were, however, welcomed by the War Office, which was grasping at any straw to stand off what it thought would be the German invasion from the skies and beaches at any moment.

After about a year, Fairbairn and Sykes were assigned to train the commandos and were ordered to train British intelligence personnel operating in the underground in France and other German-occupied countries all over Europe. From this point on, their activities were classified top secret.

In early 1942, Fairbairn was sent to the United States to help organize the combat training of the OSS, which was destined to be the U.S. counterpart of the British intelligence operations. He was, in my eyes, a crusty old bastard, about 57 years old, and I thought that made him a hell of an old man. He was about 5-feet 8-inches tall and weighed about 160 pounds.

My introduction to him gives you an idea of how ignorant some of us can be. I had met him about an hour before he spoke at what is now Camp David, where the OSS training headquarters were. I had been ordered by Colonel Donovan to be his assistant and to learn all there was to know from him, pick his brains, avail myself of his experience, add to it, and then train our own OSS people.

There was a meeting hall in one area of the camp, in front of which was a small stage. A bunch of wooden folding chairs were placed in front of the stage, and then the hierarchy of the OSS (and future chiefs of the Central Intelligence Agency), men such as Allen W. Dulles, Richard M. Helms, William E. Colby, and so on, arrived from Washington for a briefing. Fairbairn was to describe how he had done things in Shanghai, how they were now training in wartime Britain, how they were assassinating people and blowing up bridges behind the lines, and all the good stuff that happens in total war. This was all top secret stuff and was new to the Americans.

When he got into the unarmed combat discussion, he said, "Lieutenant Applegate, come out here." I walked out on the stage. He then said, "Lieutenant Applegate, I want you to attack me." I said, "You can't possibly mean that, sir," to which he replied, "I want you to attack me *for real*. That's an order."

Like many young lieutenants, I thought I knew it all. I thought to myself, "This dumb old bastard. I'll take care of him." I let out a roar and went at him with both arms wide open. The next thing I knew, I was floating through the air and, fortunately, landed on top of members of the audience in the folding chairs.

That made me very humble very suddenly, and those on whom I landed, very irritated. It was an introduction that always stuck in my mind, because at that point I decided I would listen and not have a lot of preconceived opinions that weren't backed up by experience or fact. This stuck with me always. I found, subsequently, that I usually learned something when I listened.

From my initial association with Fairbairn, I was ordered to England and worked with Sykes and others in training and operations, returning to the United States in late 1942. Shortly thereafter, I was ordered by the War Department to the U.S. Army Military Intelligence Training Center (MITC) at Camp Ritchie, Maryland, where I organized the Combat Section for training in the same subjects I had specialized in at the OSS. This was almost 55 years ago.

We trained more than 10,000 men at the center in point shooting and all phases of close combat. They were from all ethnic backgrounds, of all statures, with all sizes of hands, and with various motivations. Basically, it wasn't their ability to fight, but rather their language background, or their knowledge of enemy-occupied areas, that brought them to the center. This was also true in the OSS (we had people with specific knowledge and skills coming through). Many times we were given only a few hours to train them with and without firearms, but the Fairbairn-Sykes-OSS-MITC techniques worked.

I had 28 officers and an equal number of noncoms. We kept refining what the British had been doing but were still using the same basic principles. I sent not only trainees into combat, but instructors who would go out on special assignments and then come back and tell us if our techniques in shooting, knife fighting, strangling, or whatever worked. If a technique didn't work, we threw it out. Remember, also, that we were getting quite an education from our own students, who were from diverse backgrounds and social levels. We had MITC graduates going out into military units and then coming back to report. We had battle reports from our own intelligence and that of our enemies and allies. So point shooting, contrary to some of the so-called gun gurus, is not an untested, untried theory.

In 1943, I wrote *Kill or Get Killed*, in which I covered most of the point-shooting techniques that we had developed early in the war. There were no complete manuals available for training in close combat with and without weapons. *Kill or Get Killed* is now in its fifth edition (eighth printing) and is available from Paladin Press.

In 1944, during my Camp Ritchie training experience, the U.S. War Department authorized the filming of an official training film to be titled *Combat Use of the Handgun* (FB152). This black and white, 16mm training film was professionally made by a Hollywood crew that had been assigned to the U.S. Army and sent to me at Camp Ritchie, and the film was classified restricted. As late as 1951, you could go to the army film library and pull it off the shelf. That's the last time I saw it. I went off to Mexico and a few other places and did not think much about it for the next 40 years.

With this as background, let us look specifically at what was taught during the war from available documents in the development and techniques of close combat.[1]

—Rex Applegate

MAJ. CHUCK MELSON'S THOUGHTS

When America entered World War II, the public and armed services were bombarded with ready-made prescriptions for close combat from many self-proclaimed "Tough Guys."[2] Put forward was an exaggerated and dramatic view of the skills needed from a variety of sources, ranging from the academic to the absurd. Even the U.S. Army's manual on the subject (FM 21-150, 30 June 1942) was criticized for being limited to unarmed self-defense. One army officer complained, "I have yet to see a book on hand-to-hand, do-it-or-die fighting that tells you to shoot first, then fight some other way if you have to. I've seen no get-tougher-and-tougher book, either, that even tells you to attack. They are all manuals of the defense. The enemy jumps you first."[3] Not offered was a realistic approach to close combat that emphasized the practical options a fighter could use when restraint techniques were not enough.

Close combat occurs at the distance of a bayoneted rifle and is measured in feet rather than yards. At an earlier time in North America, Maj. Robert Rogers gave his men good counsel on the subject: "Don't stand up when the enemy's coming against you. Kneel down, lie down, hide behind a tree . . . Let the enemy come till he's almost close enough to touch. Then let him have it and jump out and finish him with your hatchet."[4] As can be seen, then as now, close-quarters combat involved the use of gun, knife, club, and hand-to-hand fighting skills.

According to former British paratrooper and martial arts instructor Jim Shortt, military close-quarter battle "commences with the rifle. If this fails, the traditional back-up is the pistol. Out of ammunition, or with no other alternative, the soldier falls back on edged weapons, starting with the bayonet. A step down from the bayonet in edged weapons is the entrenching tool used as a cleaver. Down again from this is the utility knife. If for one reason or another the soldier finds himself unarmed, he turns to unarmed combat—the object of which is to get armed as quickly as possible."[5]

Specially employed personnel had a somewhat different perspective than soldiers on what was needed when the enemy was up close and personal, and they relied on the instruction as presented in this book. The World War II fighting methods of Lt. Col. William E. Fairbairn and Maj. Eric A. Sykes, both of Great Britain, and Lt. Col. Rex Applegate have all been cited by current authorities but seldom taught, for the same reason these methods were not adapted by mainstream athletic, martial arts, and marksmanship fraternities: they were too pragmatic for routine doctrine, institutions, and instructors. Here are their classic solutions to close combat. Try them; they work.

Another viewpoint on this comes from the enemy—the Germans, Italians, and Japanese—who were being eliminated the old fashioned way—one by one. By 1942, the Germans felt that "the English and the Fighting French, probably also the Americans and Russians, have formed with their 'commandos' a new sort of troops, even indeed a new arm of the service. It consists ideally of personnel especially suited to their profession given a thorough and systematic special training, and armed with special equipment for sabotage and raids."[6]

By war's end, even the Germans emphasized close-quarter battle (*nahkampf*) for their own special forces: "Often the whole mission depends upon overcoming the enemy silently, during an unexpected encounter or when disposing of a guard."[7]

On the other side of the world, the Japanese also reached conclusions about the Allied close-combat style. In 1945, after sparring with captured "Z" Special Unit personnel, interrogators concluded they were much stronger than expected when paired against soldiers with some training in judo, and they would be difficult to defeat by the less skilled. A *Kempeitai* member holding a judo black belt stated, "I am astonished you are all so good after such a short study during your war training of what is a difficult Japanese art." [8]

Rex Applegate summed things up this way:

> Since the time of the caveman, techniques of personal combat have been in the process of evolution. There are many methods and systems of personal combat. The methods of teaching them are equally varied. Some good, some bad, some practical, others nonpractical. This [book] does not, and could not, cover all methods. It is a compilation of the most practical methods known to the writer, methods that have been developed and used during and after World War II by our own police and military, those of our allies and even our enemies.[9]

What follows is compiled from official and long unavailable wartime references used by Applegate, Fairbairn, and Sykes. The documents were selected from Colonel Applegate's own files accumulated during a half century of research and practice in military and police sciences. These were used to teach this style of close combat during World War II, from techniques developed at the time by the Special Operations Executive (SOE), OSS, and the U.S. Army Military Intelligence Division (MID), including combat shooting and armed and unarmed offense.

The book is divided into eight chapters, arranged to follow the development of the techniques rather than the sequence in which documents were obtained, while dates given are those found on the documents themselves. Documents are presented as written with only minor revisions for spelling, typographic errors, and clarity. Military ranks used for personnel are generally the highest held during the war, except for the three protagonists. Historical context is briefly outlined to identify players, institutions, and references.

Chapters 1 through 3 deal with the background and development of training in the SOE and OSS from 1940 to 1942. Chapters 4 through 7, the bulk of the text, address the specific application at the MID's training center from 1943 to 1945. Since then, there has been continued evolution, though not necessarily progress, for a variety of reasons, which is indicated in Chapter 8. Because of this, it is worthwhile to review the original lesson plans for instruction from the horse's mouth, so to speak. A bibliography lists works by Applegate, Fairbairn, and Sykes for further reading. Keep in mind that it is the science that is documented, not the art, particularly with regard to the impact of individual instruction and experience.[10]

—Chuck Melson

ENDNOTES

1. Variations of Applegate's introduction and postscript were previously presented in *Handguns*, *Tactics*, and at TREXPO West.
2. Wayne Whittaker, "Tough Guys," *Popular Mechanics*, February 1943, 40-45.
3. Lt. Col. John V. Grombach, "Kill or Get Killed," *Infantry Journal*, February 1943, 46-48.
4. Standing Orders, Rogers' Rangers, 1757 (U.S. Army Infantry School, Ft. Benning, GA).
5. Jim Shortt, "Bodyguard Close Quarter Battle," *Combat & Militaria*, September 1995, 53, 55-56.
6. "German Proposals for the Improvement of Defenses Against British Commando Raids," 2 August 1942 (Washington, DC: M2, HQMC, 26 December 1942).
7. *Werwolf, Winke fur Jagdeinheiten* (Berlin: 1945), 27.
8. Prisoners were from Operation Rimau; Ronald McKie, *The Heroes* (New York: Harcourt Brace, 1960), 276.
9. Rex Applegate, *Kill or Get Killed* (Boulder, CO: Paladin Press, 1976), xi.
10. I spent 25 years as a U.S. Marine and served in special-purpose units analogous to those dealt with in this narrative. I recall a family friend, William Durney, who told me that in World War II, as a U.S. Army officer in North Africa and Persia, he had been taught to shoot the Colt .45 by aiming it at the target "just like pointing your finger." C.M.

CHAPTER ONE

WHAT DID YOU DO IN THE WAR?

A look at the record establishes Rex Applegate's unique position in the development of close-combat skills in the last half of the 20th century. This ranged from service with the OSS, SOE, and MID in World War II to continued contributions to today's security needs.

He got there with a Western Union telegram on 16 March 1942. On that date, 2d Lt. Rex Applegate,

Rex Applegate in his gun room at home in Scottsburg, Oregon. His 55 years plus of close-combat experience began with the material documented in this book. (Photo courtesy of Cathy Cheney.)

024589, Infantry, was relieved from duty at Headquarters Western Defense Command and Fourth Army, The Presidio, San Francisco, and "assigned to Office Coordinator of Information, Washington, D.C."[1] The gregarious, 6-foot 3-inch, 230-pound, 28-year-old officer had no idea what this meant, and even the colonel he worked for could not get the orders changed or explained. The adjutant general of Fourth Army Headquarters requested information on the COI when he asked for the orders to be revoked. The reply was, "This is an order authorized by the President of the United States,"[2] and to carry it out.

A native of the Pacific Northwest, nothing out of the ordinary in Applegate's background seemed to prepare him for the assignment that defined his wartime experience and later career. Born in Oregon on 21 June 1914, his family roots went back to the American Revolution and the pioneering trek on the Oregon Trail. A youthful marksman and hunter, his uncle was Gus Peret, a Remington-Peters exhibition shooter and professional hunter. Upon graduating from the University of Oregon in 1940 with a degree in business administration and as a member of the Reserve Officer Training Corps, Applegate was commissioned a second lieutenant in the U.S. Army Reserve and was stationed at Ft. Lewis, Washington, with the 3d Infantry Division. Because of his size, taste for contact sports, and his father's experience as a lawman, he was assigned to the 209th Military Police Company, where he was when war came to America. Soon after the 7 December 1941 Japanese air attack on Hawaii, Applegate witnessed firsthand the effects of street riots while on military police duty in Seattle, assisting civil authorities to maintain order in the wake of mob attacks on Japanese-Americans.[3]

By February 1942, Applegate was selected for a regular commission before a competitive board (which contained an officer who remembered him when the army was looking for personnel for special assignment). He transferred for further training, including the U.S. Army's Counter Intelligence Corps basic course in Chicago, Illinois. As international tensions and national readiness escalated, Applegate was then sent to San Francisco for counterintelligence duties with the G2 staff Fourth Army.[4]

Applegate during World War II at Camp Ritchie, Maryland, in one of a number of facilities used to instruct intelligence personnel in combat firing and close combat. Note the posters in the background, which figure in his book Kill or Get Killed.

Characteristic of specially trained personnel was the use of so-called point or instinctive shooting of firearms in close combat, as taught by the OSS and MITC during World War II. Legendary U.S. Marine raiders—an operational concept reflected today in the Corps' Marine Expeditionary Unit-Special Operations Capable (MEU-SOC) battle configuration—use automatic pistols in the distinctive point-shooting crouch. (Photo courtesy of the United States Marines.)

Military close combat included the use of the rifle and bayonet. American techniques were developed in conjunction with the study by MITC of enemy methods, German in this example, to take advantage of strengths and weaknesses of their systems.

This was when concerns for security were at a peak. On 2 March, Lt. Gen. John L. DeWitt, commanding the Western Defense Command, established the western states as military areas. This was just three months after Pearl Harbor, and the concerns on the West Coast were about further strikes and possible *fifth column* subversion. A problem was what to do with the large number of aliens within these states: some 58,000 Italians; 40,869 Japanese; and 23,000 Germans. The army wanted to physically remove them, although the Department of Justice did not, a position supported by the navy and Federal Bureau of Investigation. One result was that some 40,000 Japanese citizens and 70,000 Americans of Japanese descent were forced into 10 Wartime Civil Control Administration relocation camps.[5]

Armed close combat with knife or club was emphasized for those who lacked more effective weapons or needed to rely on concealed or expedient protection.

The United States was still gearing up for the war that had been going on for years in Europe, Africa, and Asia. Travel space across country was at a premium, but Applegate found that the orders cut a lot of red tape, even if he did not know what the COI represented. He recalled, "On arrival in Washington, D.C., it took me three days to locate the office from which the telegraphic orders had been sent." Eventually, a colonel in the munitions building told him where to find the COI. This was at the old public health building by the Potomac River in the Foggy Bottom neighborhood. "Within an hour of my reporting, I was being interviewed by Maj. Gen. William J. Donovan. Donovan and I believed I was among the first, or the first, regular army officers assigned."[6]

A Wall Street lawyer, Donovan had a Congressional Medal of Honor from World War I and was a former assistant attorney general of the United States before being named as coordinator of information and later director of strategic services. Archivist Lawrence H. McDonald wrote that "the U.S. Army, Navy, State Department, Federal Bureau of Investigation, Secret Service and Treasury Department all had offices for foreign intelligence, and COI was to synthesize and disseminate intelligence from all these agencies." In the end, Donovan was unable to win the support of this diverse group.[7] Donovan's main concerns in 1942 were for the rapid expansion and consolidation of his organization while still only a colonel, apparently acting at times as his own personnel officer.

Applegate found that the COI, later the OSS,[8] was involved in a number of interesting and risky ventures to help fight the war. Donovan asked Applegate what he thought about being parachuted behind Japanese lines in the Philippines, to which Applegate replied, "It doesn't sound like a very good idea to me." Donovan laughed and commented, "At least you're an honest man," and went on to the business at hand of giving Applegate a more immediate task, which was assisting in establishing schools for prospective operatives.

Some understanding of Donovan's general requirements is needed to examine this training. Cols. David K.E. Bruce and M. Preston Goodfellow had inaugurated special *activities* (the COI functional branches): secret intelligence (SI) "to obtain information outside the Western Hemisphere by secret means, principally through undercover

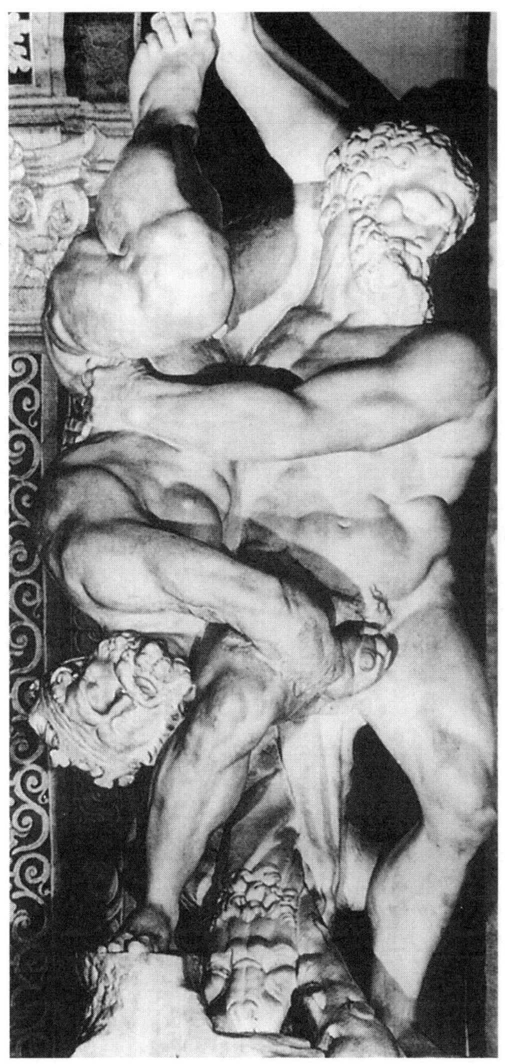

The MITC Combat Section brought its own sense of humor into play with the subject, if not good taste. Baroque sculptor Vincenzo de'Rossi depicts Hercules and Diomedes in classic close combat.

Along with firearms and armed close combat, unarmed fighting received considerable attention and included a realistic approach to the topic that debunked contemporary clichés that made claims not supported by fighting experience.

agents," special operations (SO), and subsequent operational groups (OG) "to organize and execute morale and physical subversion, including sabotage, fifth column activities and guerrilla warfare."[9] The SI personnel were civilian or military men and women engaged in espionage, sabotage, and subversion; the SO personnel were normally military individuals or small teams working with resistance and guerrilla forces; and the OG command consisted entirely of uniformed personnel organized to operate as units with guerrillas or on their own in remote areas. In time, other branches unfolded to handle research and analysis, psychological operations, and counterintelligence.

The OSS' official history comments that an initial chore "was to establish training facilities and programs which would produce spies, saboteurs, and guerrillas." It went on to summarize the problems involved in this:

> There was no precedent in America for such an undertaking and it was necessary at first to piece together various fragments of seemingly relevant knowledge from other agencies of the government, to borrow instructional techniques from the British, and to adapt certain technical aspects of orthodox military training to the probable conditions in which guerrilla units and resistance organizers might operate.[10]

While Americans in general and the COI in particular had little background with "the secrets war," the British had more recent experience to offer.[11] Donovan himself said that the British Security Coordination (BSC), led by Sir William Stephenson, "taught us all we ever knew about foreign intelligence."[12] Because of this, Applegate soon learned firsthand what the British brought to close combat after three years of war and applied it to what Americans needed in similar circumstances.

Applegate recalled: "I was given $50,000 in unvouchered funds and told by Donovan to go up to the Appalachian region near Thurmont, Maryland, and do something with it. I drove a Jeep up into an area where you are a stranger if you had not lived there for 100 years."[13] In addition to finding a training area, Donovan tasked him with finding out all there was to know about close combat "with and without weapons"[14] for use in training special operations personnel. With that, Applegate was off and running to establish a training facility.

ENDNOTES

1. Modification to Original Orders, 161808, March 1942.
2. Rex Applegate, personal communication, 20 October 1997.
3. Wiley Clapp, "Colonel Rex Applegate: American Hero," *Shooting Times*, March 1997, 70-73, 83.
4. Rex Applegate, personal communication, 12 June 1993. Lawrence H. McDonald commented that the OSS selected personnel from a pool rather than simply accepting replacements assigned by others. Over time, this exceeded 22,000 (personal communication, 24 February 1997).
5. Allan R. Bosworth, *America's Concentration Camps* (New York: W.W. Norton, 1967), 63. During 1942, the FBI seized 12,000 enemy aliens: 3,500 were released after questioning; 2,900 were paroled by the U.S. Attorney General; 1,000 were freed by hearings; and some 3,600 were interned (46). See also Audrie Girdner & Anne Lofts, *The Great Betrayal* (New York: Macmillan, 1969).
6. Rex Applegate, personal communication, 12 June 1993, 11 September 1997.
7. Lawrence H. McDonald, "The OSS: America's First National Intelligence Agency," *Special Warfare*, February 1993, 25.
8. COI was established in July 1941 and evolved into the OSS in June 1942.
9. OSS, *War Report Office of Strategic Services*, Volume I (Washington, DC: USGPO, July 1949), 70-84; OSS, Unclassified Report: "Aid to Resistance," 14 September 1945, passim; McDonald, "The OSS," 24-32.
10. OSS, *War Report* I, 231.
11. After World War II, SI and some SO functions were retained in the CIA. Defense intelligence agencies also continued to have some SI functions. Other SO and OG roles were retained in the armed forces, including those of the U.S. Army Special Forces, which modeled its structure on a combination of operational groups and Jedburgh teams resulting in the basic A Team structure. OSS, *Operational Groups Field Manual*, 25 April 1944, passim; Aaron Bank, *From OSS to Green Berets* (New York: Pocket Books, 1987), 14-15, 176-177.
12. H. Montgomery Hyde, *Secret Intelligence Agent* (New York: St. Martin's Press, 1982), 248, 255; see also Thomas F. Troy, *Wild Bill and Intrepid* (New Haven: Yale University Press, 1996), passim.
13. Rex Applegate, personal communication, 30 October 1995.
14. Rex Applegate, California Rangemasters Association, Burbank, CA, 21 March 1996.

CHAPTER TWO

WITH THE SPECIAL OPERATIONS EXECUTIVE

THE EMBATTLED ISLAND

Whereas this book focuses on events between 1940 and 1945, because of the British support of the American intelligence effort, the story really begins in Shanghai in the 1920s and 1930s. There, two of the players got their start with the SMP. William E. Fairbairn was a marksmanship and drill instructor, and later headed all training. With these responsibilities he evolved a unique system of unarmed self-defense and pistol shooting to meet the needs of the force. He also led the riot squad, an early special weapons and tactics/emergency response unit. He was joined in this by Eric A. Sykes, a firearms and ammunition salesman, who ran the police sniper element as a special policeman. Professional exchanges and articles made their developments known in the United States to law enforcement and shooting communities in the 1920s.

American soldiers, sailors, and Marines worked directly with the SMP from 1927 until 1941 and were exposed to these innovative techniques as part of their peacekeeping duties. This prewar Shanghai connection provided the start to an affinity that continued through World War II and can still be seen today.[1] The Marines' guru of close combat, Col. A.J.D. Biddle, discovered Fairbairn's "Defendu" at this time and passed it on with enthusiasm.[2] The "Three Sams" of the 4th Marines—Yeaton, Moore, Taxis—were some of those who worked with Fairbairn and Sykes (as documented by Applegate and the Yeaton brothers in *The First Commando Knives*).

With the start of fighting in Europe in 1939, hostilities erupted between the British, French, Germans, Italians, and Japanese before America's entry into the war. The British military left Shanghai as the war in Europe began, while U.S. forces remained there longer. The war in Europe was well underway in 1940 when Fairbairn and Sykes returned to the United Kingdom from the Far East. As noted earlier, both were picked up by the War Office to be army instructors, having come to the attention of the Directorate of Military Intelligence's Section GS(R), later MI(R), and Maj. Gen. Colin McVeigh Gubbins.

In May 1940, after abortive offensive efforts to Finland and Norway, a group of like-minded individuals gathered at what became the Special Training Center (STC) at Lochailort, Scotland. Included were Lord Lovat, William and David Stirling, F. Spencer Chapman, and Michael Calvert. As the program evolved, the demolition wing was run by Calvert and James M.L. Gavin, the fieldcraft

Firearms training by the SOE used the point method developed by Fairbairn and Sykes. This student courier is taught equality according to "Judge Colt." (Photo courtesy of the Imperial War Museum.)

wing was run by Lord Lovat and Chapman, and the weapons training wing included Fairbairn and Sykes.

Gubbins had already outlined the needs for special forces and operations, noting that the object was to accomplish the maximum damage in a short time and then to get away. In this, explosives of all types were "of great use," and while "the most effective weapon for a guerrilla is the submachine gun, which can be fired either from and rest or from the shoulder," and pistols were important for close range or night work, bayonets were unsuitable for guerrilla fighting: "these are only for use in shock action which should be eschewed; a dagger is much more effective and more easily concealed."[3]

The school staff went through the course first and then concentrated on the training of the independent companies, later renamed commandos, billeted at Arisaig, Glenfinnan, and Morar. Courses lasted six weeks for officers and noncommissioned officers. The June 1940 commando charter called for them all to be volunteers, trained to use "all infantry weapons" and given the techniques for "killing or capturing quickly and silently."[4] A holding wing was subsequently formed at Lochailort for commando replacements until training began at Achnacarry in later years.

Instruction was as practical as it could be: demolitions taught to blow up anything from "battleships to brigadiers." Land navigation taught how to move from point to point and back by day or night. Chapman wrote that Fairbairn and Sykes "converted a derelict cottage into a sinister revolver range where targets suddenly appeared in the half-light and had to be engaged from any position."[5] It was pointed out by one student, a conventionally trained marksman, that "though my target would be nearer, in a room, for instance, I might also have to shoot in the dark, when I could not see the sights. Or I might have to shoot fast, with no time to align the sights." [6]

Left: Techniques were presented in classroom settings and then applied in field exercises. A sand table is used to review terrain appreciation and a tactical scheme to a group of SOE pupils. (Photo courtesy of the Imperial War Museum.)

Right: Advanced preparation included radio theory, code, and ciphers. An SOE radio operator practices manual Morse on the clandestine "suit case" radio used in occupied Europe. (Photo courtesy of the Imperial War Museum.)

With the evacuation of British forces from Dunkirk, STC instructors were employed in preparations for a potential German invasion (a street-fighting school, home guards for local defense, and stay-behind parties for intelligence and sabotage) while others prepared organizations to strike into occupied territories (commandos, parachutists). This proliferation of "special" forces soon found the STC brand of close combat useful.[7]

Through July 1940, Fairbairn and Sykes busied themselves in training instructors for close combat, while at the same time preparing the police, civil defense, and home guard personnel for the threat of German invasion. This was a desperate time calling for desperate measures. While Americans provided needed small arms to the British, plans were being made to catch and kill any German and "take his weapons and ammunition and use them to kill more Germans."[8] Stressed was the use of any improvised weapon at hand, even a "carving knife, cleaver, or trowel,"[9] until better ones could be obtained. Fairbairn and Sykes made the most progress in developing training for specially employed personnel that by its nature has remained in the shadows, and they made a change from defensive to offensive methods in close combat.

On 22 July, the SOE received its charter as the "Fourth Arm" of the British services and brought together Secret Intelligence Service Section D, Military Intelligence (R), and the Department of Propaganda (EH). The SOE initially consisted of two functional branches: propaganda (SO1) and

Basic parachute, firearms, and hand-to-hand training were considered essential skills for SOE personnel. A dispatcher hooks a student up to the aircraft's parachute static line prior to a jump. (Photo courtesy of the Imperial War Museum.)

operations (SO2). The auxiliary units had been transferred from Section D to Gubbins' MI(R), and in October 1940, MI(R) itself became fully part of the SOE, though SO1 later separated to join the Political Warfare Executive (PWE).[10]

In collaboration with the long-established special branch, military and naval intelligence, secret intelligence, and security services, the SOE took over ongoing enterprises in Poland and France and established itself for operations from the United Kingdom into occupied Europe. Dealt with were collaboration with the Foreign Office, the military, covert funding, radio facilities, and recruitment. This required a steering body to select projects, which in turn were passed to intelligence, planning, and the director for operations and training. Maj. Gen. R.H. Barry wrote that concepts of operations had to be foreseen—air and sea movement, communications, supplies, sabotage, information, and escape.[11]

Gubbins noted the necessity for "blind drops"; the first operatives into occupied territory were parachuted without reception committees, and this continued "to be the most efficient means of insertion. The coasts are guarded, the waters mined, and landing agents by sea results in high losses. This requires highly specialized training in the air, for both pilots and agents." In clandestine operations there was a 3-in-100 chance of being captured and a 1-in-3 chance of being interrogated by the Germans after arrest. Network radio operators had the highest risk of all and had a useful life of only about six months before being killed or seized.

A "secret intelligence agent" goes through the "Joe" hole of a transport aircraft with remaining team members lined up behind him. (Photo courtesy of the U.S. Air Force Special Operations Command.)

All these activities depended upon trained personnel, and facilities were set up with due regard for compartmentalization. Special operations recruiting, training, dispatching, and servicing were handled in a military and largely impersonal manner. Secret intelligence processing, in contrast, was on a personal basis to make each agent feel that one individual at the home desk would do everything to support them. Training was run by the SOE's M Section, which prepared "foreigners" for specific sabotage or raids in their own countries or as "organizers and agents" to be sent to occupied countries for subversion. Courses lasted 10 weeks or longer, depending upon final responsibilities. At these, "self-defense and unarmed combat were essential for everybody, then the network organizer,

Left: An exterior view of a supply drop by a B24 Liberator bomber on transport duties in support of European resistance groups. Equipment and personnel were limited to the configuration of insertion craft. (Photo courtesy of the U.S. Air Force Special Operations Command.)

Below: Special operations personnel on an aircraft over occupied Europe late in the war. Uniformed and undercover personnel were trained for close combat in the methods covered by Fairbairn, Sykes, and Applegate. (Photo courtesy of the U.S. Air Force Special Operations Command.)

the wireless operator, or the sabotage expert could concentrate on learning his special skills."[12]

Depots provided the initial instruction where the unsuited were sorted out, and then it was on to paramilitary or finishing schools. This evolved to three levels of instruction at a variety of special training schools (STS): paramilitary at the Group A schools, finishing at Group B, and operational preparations at Group C establishments. In Great Britain, new arrivals were given a tentative assignment with some SOE branch or section, then went on to a reception center. While there, their security was checked and they received a limited amount of introductory training in the use and care of small arms, map and compass reading, fieldcraft, and close combat. An obstacle course provided conditioning, and facilities for recreation included ball fields and a swimming pool.

After a security clearance was obtained, recruits underwent a four-day assessment board where motivation, intelligence, aptitude, attitude, stability, discipline, leadership, self-confidence, physical coordination, and stamina were tested. The results were reported to a training officer who passed them on to the section concerned. Those rejected were transferred to other branches, given routine operational duties within the organization, or released. Motivation was considered the first and most indispensable qualification for success.

When the results of the assessment were studied, arrangements were made for further training, depending upon the recruit's background and ultimate assignment. Agents usually went first to a five-week paramilitary course. This included instruction in weapons, unarmed combat, demolitions, guerrilla warfare, basic communications, intelligence, small boats, and organizing resistance groups.

Upon completion of this course, agents went to parachute school, which lasted from four to six days, depending upon the weather. In this short time they conducted ground exercises and made four practice jumps: three by day and one by night. This was followed by a three-week finishing course in all phases of agent activities, followed by a four-day field problem. This final test allowed instructors and section representatives to evaluate what the student had learned and his ability to use it.

Most students attended one or more specialized schools operated by the SOE. Wireless training school at Thames Park was the most demanding. Here prospective wireless-telegraph operators learned code, ciphers, and radio theory. Other courses were on foreign weapons, mines and booby traps, industrial sabotage, propaganda, street fighting, reception committees, and aircraft landing and pickup.

After an agent had completed a required combination of courses, he went to an operational holding area to remain until leaving for enemy territory. At this point he became available to his parent section, which began to prepare him for cover, documentation, and infiltration. An agent could then focus on a cover story, briefings, and dispatch to "the field."[13]

Expansion of needed training and support facilities eventually occurred in the Middle East, Southeast Asia, and North America. These overseas schools combined similar substance at one location. By the end of 1940, the SOE set up a school similar to Lochailort in the Far East to train New Zealand and Australian independent companies to conduct both guerrilla and commando actions. Emphasized were guerrilla tactics and skills, demolitions, use of enemy weapons, map reading, night navigation, agent circuit operations, intelligence, sabotage, escape and evasion, counterespionage, ambushes, security, use of couriers, and hand-to-hand combat. Instructor Michael Calvert recalled he used jujutsu to "kill, or be killed" as Fairbairn and Sykes' style close combat went to this theatre.[14]

Fairbairn and Sykes' methods developed into the form most are familiar with: basic close-combat instruction divided into two parts of roughly eight hours each for silent killing and battle firing. Fairbairn outlined the system to students and the justification for using such drastic methods.

> The methods in which you are going to be instructed have stood the practical test of actual use over a number of years in one of the toughest cities in the world: Shanghai. Naturally under such conditions as exist in present-day warfare, one must have no scruples as to the way and manner in which it is intended to kill an opponent. The justification for using any foul method in accomplishing this end does not, when one stops to think what has already been used against us, have to be considered.[15]

Fairbairn felt that British and American soldiers had "a natural repugnance to this kind of fighting. But when he realizes that the enemy will show him no mercy, and that the methods he is learning work, he soon overcomes it."[16] He concluded that "in modern warfare, the job is more drastic. You're interested only in disabling or killing your enemy. That's why I teach what I call 'Gutter Fighting.' There's no fair play; no rules except one: kill or be killed."[17]

Instruction expanded to include shoulder weapons and two edged weapons, the so-called Fairbairn and Sykes No. 1 and No. 2 fighting knives.

The famed Fairbairn-Sykes fighting knife was first produced by the Wilkinson Sword Company in January 1941 for distribution to special training center alumni, along with the bolo-like smatchet.

The fighting knife was a close-combat weapon, excellent for stealthy attack but not designed for all-purpose use, whereas the smatchet was designed to fill both fighting and utility roles. Edged weapons supplemented firearms or were used by the operator as his sole means of defense or offense. The knife was double-edged and could be used either for penetration or cutting. It was carried in a special scabbard designed to be worn high or low on the belt, angled into any position, or fitted snugly to any part of the body, with ready access with either hand. Each student was "provided with a fighting knife, which is used by the commandos with particular effectiveness," reported American observers.[18]

Fairbairn advised:

> There isn't a man in the world who isn't afraid of a bright, gleaming knife flashing menacingly near him. On the other hand the feel of a sharp knife in your hand gives you confidence and the aggressive spirit, for every good knife man knows that 1/6 of an inch of blade reduces anybody in the world to his own size, weight and strength.

Fairbairn added:

> You'd be surprised how much confidence it gives you and how quickly you can acquire a fighting skill with it.[19]

An American trained in this method concluded that the Fairbairn-Sykes knife:

> . . . is not for all people, places, and seasons. It is no good as a survival tool—unless killing people means survival. The Fairbairn was magnificently designed for the job it was intended to do—by someone who knew what he was doing.[20]

An early STC instructor, William Pilkington, concluded that "any knife is a combat knife, if you are in combat."[21]

Another weapon was the club-like "spring cosh," which was borrowed from the German security police (*Sicherheitspolizei*). Author Michael Crichton wrote about the other illicit use for this, the underworld's "eel-skins" and coshes:

> The earliest eel-skins were sausage-like canvas bags filled with sand, which rampsmen and gonophs—muggers and thieves—could carry up their sleeves until the time came to wield them on their victims. Later, eel-skins were filled with lead shot, and they served the same purpose. A "neddy" was a cudgel, sometimes a simple iron bar, sometimes a bar with a knob at one end. The "sack" was a two-pound iron shot placed in a strong stocking. A "whippler" was a shot with an attached cord, and was used to disable a victim head on; the attacker held the shot in his hand and flung it at the victim's face, like a horrible yo-yo. A few blows from these weapons were certain to take the starch out of any quarry, and the robbery proceeded without further resistance.[22]

Though trained with the pistol, most operatives did not carry firearms or any special devices to avoid routine searches, except when vulnerable, such as during air drops, pickups, or when transmitting. Emphasized in training was that it was better to put a German in the hospital: "That ties up other Germans. A dead one is buried and out of the way." They learned that "a knife should be used delicately as a paint brush"; if unarmed, use the heel of the hand upward to smash an assailant's jaw or "knee him in the groin." Taught were leverage, vulnerable parts of the body, and that you've "always got a weapon in your pockets: a nail file, a pin, a fountain pen."[23] One agent concluded that Fairbairn and Sykes gave them "more and more self-confidence, which gradually grew into a sense of physical power and superiority that few men ever acquire." By the time he finished training, he "would have willingly tackled any man, whatever his strength, size or ability."[24]

The realities of underground warfare were perhaps too grim for a cheerfully dogmatic approach to close combat, which appealed more to those who fought in uniform and in groups. For underground agents, it was more intensely personal.

> It was over. The moment we had discussed so often—the moment of my arrest. I was in the hands of the Gestapo. During many long hours of reflection I had come to one decision: not to let them get me alive. Firstly because in our group we had all decided never to let ourselves be taken. And then again, I had another reason—I was afraid. Afraid of talking under torture. I raised my arms a little. Then, with my right elbow bent and holding my right wrist in my left hand, I aimed a blow at his solar plexus, with all the force of pent-up tension suddenly released increasing my strength tenfold. He went over backwards. I jumped over him and began to run.[25]

This French courier got just far enough away to be shot and spent the rest of the war as a concentration camp prisoner.

Fairbairn and Sykes rotated on a circuit of instruction at the various discreet training facilities through 1942. By then they were at their peak of utility in the United Kingdom. That year, publishing made the Shanghai approach to close combat available to a wider audience. Fairbairn marketed *All-In Fighting* (released as *Get Tough!* in the United States) on hand-to-hand fighting, which was criticized by Sykes, who felt it was too complex and police oriented.[26] Fairbairn knew the limitations of published instruction and that appropriate techniques had been left out in favor of "those that could be mastered in a few lessons,"[27] but it was better to master 12 or so methods than be slightly familiar with a greater number.

Collaborative pistol-shooting concepts were spread by *Shooting to Live*, recognized by Applegate as:

> . . . the first written manual to surface in the field of combat pistol shooting. Its principles and techniques were expanded and modified upon to fit American needs in the various editions of the writer's own text. Many present-day U.S. military and law enforcement combat handgun techniques can be traced to this book.[28]

But Fairbairn and Sykes were also at an impasse for continued development and usefulness of their style of close combat. Neither were part of the ensconced physical training and marksmanship bureaucracy of instructors for whom range drill, bayonet, and sword fighting still held sway. Shanghai methods had been incorporated into official doctrine for both armed and unarmed defense, but were moderated to fit existing programs.[29] There was even some aversion to the whole approach by those who felt that unarmed combat "seemed to consist of low-life gangster fighting, and to include biting, gouging and unpleasantly directed kicking." Chindit commander Bernard Fergusson called it "barbaric, and forbade it." We've only fought decently in the British Army and I don't see why we should change now."[30] In fairness, another Chindit commander with broad experience felt that he "was a trained soldier, taught how to kill with a gun, or a bomb, or a bayonet or even a knife" and used it himself in one instance.[31]

Some hard feelings existed between Fairbairn and Sykes about recognition and Pilkington quoted Sykes as saying, "We seem to be serving a one-man army, everything we invent or create he channels into his own record."[32] Fairbairn was restless, in part because of being kept out of combat, with his son John a prisoner of the Japanese and daughter Dorothea with the SOE.

It was the Americans who provided an outlet from which to perfect and promote techniques. The British and American approaches to temperament were fundamentally different. These differences were seen in clandestine methodology, including the approach to close combat. The British thought a love of productive peril or a certain flamboyance was acceptable if there was moral force behind it. This was informal and imprecise. That Fairbairn was an outsider by birth, education, and class—though overcome in colonial service—was not relevant to the Americans with whom he worked next. The Americans required Fairbairn's definitive approach for industrial-modeled mass-production methods and had a respect for the self-made individual.[33]

While Sykes remained in the United Kingdom, Fairbairn arrived in Canada in March 1942 as an instructor at SOE STS-103 on Lake Ontario's north shore near Oshawa. The camp provided a location for the training of Commonwealth subjects before being sent overseas and for the instruction of selected Americans from various civil and military organizations.

H. Montgomery Hyde, a member of the intelligence service and BSC, described the STS-103 curriculum as follows:

> The new recruit was taught the importance of accurate observation; and his own powers of observation were frequently put to practical test by moving or removing objects in his room. He was taught how to shadow a man and how to escape surveillance himself; how to creep up behind an armed sentry and kill him instantly without noise; and how to evade capture by blinding his assailant with a box of matches. In the course of unarmed combat he learned many "holds" whose use would enable him to break an adversary's arm or leg, to knock him unconscious or to kill him outright. He was also given weapons training. He learned to handle a tommygun and to use several types of revolvers and automatic pistols, firing from a crouching position either in daylight or darkness. He was instructed in the dexterous use of a knife, which could kill swiftly and silently if driven upwards just below the ribs.[34]

Explosives, sabotage, codes and ciphers, communications, and interrogation were taught. Exercises took place on how to use aircraft and parachutes, evade local police, and penetrate security areas.

One anecdote of Fairbairn's Canadian stay was told by Sir William Stephenson, whose New York-based BSC office represented British intelligence concerns in North America. The exercise involved disposal of a "tail," and required a student-operator to be given a loaded pistol and for him to break into a hotel room and shoot the occupant, played by Fairbairn, "in cold blood."[35]

Hyde recalled that STS-103 provided the OSS with "all its initial instructors, books, and equipment." Gubbins commented that with America's entry into the war, "we took into our schools here in Britain and in Canada a number of [Donovan's] officers, and also lent him senior instructors to start his own schools in America."[36] Arranged through the BSC, these included Capt.

The present Commando Memorial, Spean Bridge, Scotland, in the Fort William and Arisaig region where training was conducted by British Combined Operations and SOE schools. (Melson collection.)

W.E. Fairbairn and Maj. R.M. Brooker from the SOE, and Lt. Cdr. H.G.A. Woolley from Combined Operations. Fairbairn moved with his expertise to the United States in April 1942 as part of a form of reverse "lend lease" and was assigned to the OSS' schools and training.[37]

These first documents reflect training developed by Fairbairn and Sykes in the United Kingdom, and indicate the basic close combat brought to the United States and used by Applegate and other Americans who went through the various British intelligence or special operations programs.

SEQUENCE OF INSTRUCTION—REVOLVER AND AUTOMATIC
[Document 1; 1942]

1. Introduction to the weapon.
   a. Name.
   b. Calibre.
2. How to prove.
3. How to load and unload.
4. Introduction to method of firing.
   a. Weapon of attack.
   b. Short range. Close quarter with enemy.
   c. Extreme speed. Firing from any position and in any kind of light.
   d. Visualize the circumstances, e.g., enemy-occupied house.
      1) Never upright—crouch.
      2) No time to adopt special stance.
      3) No time to use sights.
   e. Only method of firing under these circumstances is by instinctive pointing—explain and demonstrate.
5. Holding.
   a. Convulsive grip.
   b. Trigger finger inside trigger guard.
   c. Thumb on same plane as barrel, if possible.
6. Firing position: Demonstrate in detail recalling points emphasized in introduction.
7. Practice squad.
8. Explain the purpose of the large recruit target.
9. Explain reasons for two shots at every target.
10. Demonstrate how to turn to targets right and left after firing first practice.

Points to Emphasize. General.

1. Shoot to kill.
2. Fastest man wins—Speed.
3. Speed with sufficient accuracy to hit target in vital part.
4. Position must be natural and balanced. Firer alert and concentrating ready to move quickly in any direction.
5. Do not handle student more than is necessary.

Practice.

1. Recruit Target.
2. Moving Targets.
3. Moving Targets in dark.
4. Standing Targets. Firing:
   a. From behind cover.
   b. Long range—standing position, i.e., feet astride—push and pull.
5. Demonstrate:
   a. Prone position.
   b. Firing down.
   c. Firing up.

SEQUENCE OF INSTRUCTION—TOMMYGUN

1. Introduction to the gun.
   a. Name.
   b. Caliber.
   c. Rate of fire.
   d. Weight.
   e. Types of magazine.
2. Primarily close range weapon. Particularly suitable for street fighting, etc. Fired from hip up to 15 yards. Shoulder standing to 50 yards. Maximum effective range 175 yards, firing from the shoulder in prone position.
3. Manipulation of gun.
   a. How to prove.
   b. To load:
      1. Locking handle forward.
      2. Position of change lever and safety catch.
      3. Insert 20 shot magazine.
   c. Explain why gun is safe.
   d. Practice squad.
   e. Unloading.

f. Explain reason change lever at full automatic. Trigger tapping.

4. Demonstrate with detail firing from hip. Show alternative positions.

5. Practice squad.

6. Show how to turn targets left and right after firing first practice.

Points to Emphasize. General.

1. Speed.

2. Balanced and natural position—body pressed forward—aggressiveness and determination.

3. Left elbow well under gun.

Practice.

1. Ten single shots at Recruit Target.

2. Five rounds at Moving Targets.

3. Five rounds at Moving Targets in dark.

4. Further practice at Moving Targets introducing movement and building up speed.

SILENT KILLING
[Document 2; 30 June 1942]

For the use of instructors.

A course of instruction designed to teach how to fight and kill without the use of firearms. Since the course includes the use of the knife, the term "unarmed combat" would not be technically correct. "Silent killing" has been chosen, therefore, as a more accurate description.

When commencing the course with a class of untrained students, the instructor should make a short introduction, not necessarily in the same words but to the same effect as the following:

This system of combat is designed for use when you have lost your firearms, which is something you should not do, or when the use of firearms is undesirable for fear of raising an alarm.

At some time or other, most of you, probably, have been taught at least the rudiments of boxing, under the Queensberry [John Sholto Douglas, 8th Marquis and Earl] rules. That training was useful because it taught you to think and move quickly and how to hit hard. The Queensberry rules enumerate, under the heading of "fouls," some good targets which the boxer is not trained to defend.

This, however, is WAR, not sport. Your aim is to kill your opponent as quickly as possible. A prisoner is generally a handicap and a source of danger, particularly if you are without weapons. So forget the Queensberry rules; forget the term "foul methods." That may sound cruel but it is still more cruel to take longer than necessary to kill your opponent. "Foul methods" so-called, help you kill quickly. Attack your opponent's weakest points, therefore. He will attack yours if he gets a chance.

There have been many famous boxers and wrestlers who time after time have won their contests with their favorite blows or holds. The reason is that they had so perfected these particular blows and holds that few could withstand them. The same applies to you. If you will take the trouble to perfect one method of attack, you will be far more formidable than if you only become fairly good at all the methods which you will be shown.

Since this course of instruction is designed to teach you to kill, it will be plain to you that its methods are dangerous. Your object here is to learn, not to damage, and you will get no credit if you break your sparring partner's neck, for example. In learning and practicing, you will avoid, therefore, taking any risks of that kind. The submission signal (the two taps, on your body or your partner's or on the floor) must never be disregarded. It is the signal to stop instantly, and that is a rule which must never be broken.

Note. Dummies are an invaluable aid to instruction in the various blows. They are essential to the practice of Section 4. The instructor should have half-a-dozen in readiness, therefore, beforehand.

The Course. Is divided, as far as the average student is concerned, into six progressive sections. This arrangement is to be regarded, however, as elastic. Depending on such considerations as time available, progress made by students or their standard of knowledge, there is no reason, for example, why two or more sections should not be amalgamated. Again, if at the later stages the instructor thinks it necessary, in order to relieve the tedium of constant repetition, he may show at his discretion, a selection of the holds, etc., listed under "Various" at the end of this syllabus. He should keep steadily in mind, however, that students whose time is limited are only apt to get confused if shown too much. It must not be lost sight of that the primary object of the instruction is to make them attack-minded, and dangerously so.

Section 1.

Blows with the side of the hand. Explain that the most deadly blows without the aid of weapons are those with the side of the hand. To deliver them effectively, the fingers must be together, thumb up, and the whole hand tensed. The blow is struck with the side of the hand, all the force being concentrated in one small area, i.e., approximately half-way between the base of the little finger and the wrist joint, or where the hand is the broadest. If striking sideways, the back of the hand must be uppermost. No force can be obtained if the palm is uppermost.

Explain that with these blows, it is possible to kill, temporarily paralyse, break bones or badly hurt, depending upon the part of the body that is struck. The effect of these blows is obtained by the speed with which they are delivered rather than by the weight behind them. They can be made from almost any position, whether the striker is on balance or not, and thus can be delivered more quickly than any other blow.

Having explained the blows, the instructor should demonstrate them on the dummies and get the students to practice after him. His main point here is to bring out the speed of the blows and to see that students deliver them correctly.

Students should now be shown where to strike, as follows, explaining the effect on each particular point:

1. On the back of the neck, immediately on either side of the spine.
2. From the bridge of the nose to the base of the throat.
3. On either side of the head and throat, from base of the throat to the temple area.
4. On the upper arm.
5. On the forearm.
6. The kidney region.

Students should practice on the dummies again, keeping in mind the vulnerable points listed above. Strike with either hand.

Section 2.

Other blows.

How to kick. As a general rule, kick with the side of the foot and, unless you possess unusually good footwork and balance, don't kick above knee height. Never kick to the foremost unless your opponent has both hands occupied. In that case, it is safe to kick to the fork. Once an opponent is down, kill by kicking the side or back of the head (<u>not</u> the top of the head).

<u>The boxing blows.</u>

<u>The open-hand chin-jab</u>, fingers held back and apart ready to follow up to the eyes. Utilize the occasion to obtain some improvement in footwork, explaining that the body must be properly positioned in order to obtain telling effect from either boxing blows or the open-hand chin-jab. Explain, too, that neither can secure more than a knockout, which should be followed up instantly by a killing attack.

<u>Use of the knee</u>, often in simultaneous combination with other attacks (e.g., with the chin-jab). Show how, while being used for attack, it is an excellent guard for one's self.

<u>Use of the head and elbows</u>, for attack when the opponent is not in position for more effective blows.

<u>Finger-tip jabs</u>, to solar plexus, base of throat, or eyes, when nothing more effective can be done.

All the blows listed should be practiced now on the dummies.

Conclude this section by telling students, as emphatically as possible:

1. That they should never go to ground if they can help it. If they have to, they should get up again as soon as they can. While a man is killing his opponent on the ground, the opponent's friends could walk up and kick his brains out. Again, while on the ground, it is difficult to go on attacking.

2. That if their knowledge of the subject is confined to the contents of Sections 1 and 2, they will have made themselves extremely dangerous, even to highly trained adversaries, <u>if only they will attack first and keep on attacking</u>. Don't stop because the opponent is crippled. If you have broken his arm, for instance, that is only of value because it is then easier to kill him.

<u>Section 3.</u>

<u>Releases from holds.</u>

Explain first that, in general, nobody should be so slow in wits or body as to allow someone else to get a hold on him. In case of misfortune, however, show how to effect release from:

A wrist hold, taken with one hand.
A wrist hold, taken with two hands.
A throat hold, taken with one hand.
A throat hold, taken with two hands.

Show here how, instead of the customary wrist-and-elbow release or one of its variants, it is far simpler, quicker, and more effective to attack, e.g., knee to the fork and fingers to the eyes, simultaneously.

A hair hold from behind.

A body hold, from front or rear, arms free and arms pinioned.

Police or "come-along" holds.

The whole idea of releasing yourself from a hold is to enable you to attack and kill your adversary. Whenever possible, the disengaging movement should form the commencement of the attack. In any case, there must be an effective and instant follow-up attack after every release. The instructor should demonstrate most carefully every detail to do this section and then insist on students practicing until the mode of release and the subsequent attack becomes a matter of instinct, to be carried out at lightning speed. The instructor should emphasize the importance of footwork and, where necessary, try to improve it.

Section 4.

Crowd fighting.

One cannot always choose when one will fight and it may sometimes happen that one is faced with several opponents at once. On such occasions, unarmed yourself, your object is not so much to kill your opponents as to get quickly away from them so that you do not get killed. Pride is expensive if it entails defeat and death.

To escape from circumstances like these, a special technique is necessary.

For this technique, balance is essential and the instructor should now demonstrate how to keep on balance when swift movement is necessary in kicking while standing on one foot. Students can be paired off and standing on one foot, arms folded, they should try to kick each other off balance whilst maintaining their own balance.

Once this is mastered, it should be explained that, surrounded by a crowd, your only chance of escape lies in continual movement. This is so because, after you have taken up a new position, it requires a second for your opponent to turn and balance before he is able to strike you with any force. If one moves at least three feet in each second, there is obviously little chance of an opponent scoring an effective hit on you. At the same time, by the use of the blows previously learned, you will be able to do considerable damage while you are moving.

Note. 1. In addition to forward, backward, and lateral movement, move also at different levels, sometimes with the knees very much bent. It also helps, if done at speed, to bewilder your adversaries.

2. Of necessity, there will be little room for movement, so make room by moving against one opponent after another, attacking as you do so. Point out the value of the balance and footwork, which the students should have practiced at the beginning of this section.

The information contained in the two above notes should suffice to prepare students for the actual practice, which is now outlined.

Five or six actual dummies should be suspended in a confined space; a boxing ring would answer the purpose. One student at a time should enter the ring, and with all the speed with which he is capable, he should then attack the dummies at random, using every kind of blow with hand, foot, knee, elbow and head, from any position.

Practice is very exhausting and cannot be kept up for more than a minute.

The instructor must carefully watch for faults so that he can give advice afterwards.

Before the student tires, he should be told to leave the ring and he will do so at speed, exactly as if he were actually making an escape.

To derive the maximum benefit from this exercise it should be first done both by the instructor and the students in slow time, paying careful attention to footwork.

It should then be followed by many short periods in the ring and only an occasional longer one. It must always be remembered that the aim is to get out of the place and not to fight any longer than necessary.

Section 5.

Knife fighting.

The knife is a silent and deadly weapon that is easily concealed and against which, in the hands of an expert, there is no sure defense, except firearms or by running like hell.

Students should be taught how to hold a knife, how to pass it from one hand to another, to thrust, and how to use the disengaged hand to feint and parry. It is unnecessary to be ambidextrous to be able to use the knife with either hand.

After practicing these, show them the vulnerable points, slashes permissible and how to make an opening for a thrust by flinging a hat, a handful of gravel or some other object in an adversary's face. Stress the value of surprise, showing the opportunities for it.

Explain the various methods of carrying a knife, and the value of a really sharp point and edge, the latter being to prevent the knife from being seized as much as for slashing.

After the foregoing has been fully mastered, explain the possible defenses against the knife, such as the parries, kicks, use of a chair, and of a steel helmet as a shield.

Section 6.

Useful aids, for special needs and occasions.

1. Killing a sentry, if you are armed with a knife:

a. From the rear; striking on the side of the neck, transferring the hand over mouth and nostrils, thrusting simultaneously with the knife in the other hand in kidney or abdominal area.

b. From the rear; strike on side of neck, turn the chin up, thrust with knife in the other hand either in rear of the carotid artery, cutting forward, or in front of the carotid, cutting backwards.

2. Killing a sentry, if you are unarmed:

a. From the rear; simultaneous blows across the throat, with the forearm, and in the small of the back, with a clenched fist. With the hand, which is struck in small of back, cover mouth and nostrils.

Slip around to the front, gripping around his neck with the arm, which is struck across the throat, trip over one leg and take him heavily to the ground. Lie diagonally across him and transfer the hands swiftly to the stranglehold, one thumb on either side of the windpipe and two fingers of each hand on either side of the spine.

b. From the rear; use the head hold and take him down onto your thigh. Keeping the hold in the proper manner, sit down instantly with legs stretched out in front of you. (Instructors will see to it that their students do not sit down when learning or practicing on each other, while keeping the head hold.) Depending on respective sizes of sentry and attacker, it is sometimes an advantage for the latter to get his knee in the small of the sentry's back.

It goes without saying that in all four methods of killing a sentry, extreme speed is essential. There are other methods of killing a sentry, but the four described seem to be the simplest to teach and the most effective in operation.

3. Spinal dislocator, opponent sitting. [Approach from rear. With your left hand under the sentry's chin, drag his head back completely under your right armpit. Drop your left hand on his left shoulder and, passing your right arm across the back of his neck, grip your left wrist from above. The finishing touch is a quick snap upwards and backwards. A very dangerous hold and requires great care in practicing.]

4. Disarming, if held up with a pistol: [Explain first that only a fool would hold you up with his pistol within reach of your hands. Nevertheless, it is plainly evident that there are still a lot of such fools about and if you did not know how to deal with them it would be you who would feel a fool.]

a. From the front.

[Method A. Hands up, well above your head and wide apart. Don't look at the pistol. Bring your right hand down smartly onto his wrist, gripping it firmly with your thumb preferably above. Accompany the movement by a half turn to your left. Simultaneously, your left hand grips the pistol barrel from underneath and presses the pistol backwards. Note that while the pistol is being pressed backwards, its barrel should be parallel with the ground. This will break your opponent's trigger finger and give you possession of the pistol. Turning half right, attack with your foot or knee to the fork, open-hand chin-jab, butt with the top of your head, or anything calculated to knock your opponent out. Each movement has been described separately but, in actual practice, the movements should be performed so quickly that they appear to be almost one.

Method B. Hands up, well above your head and wide apart. Don't look at the pistol. Bring your left hand down smartly onto his wrist, gripping it firmly with your thumb preferably above. Accompany the movement by a half turn to your right. Simultaneously, your right hand grips the pistol barrel from underneath and presses the pistol upwards, backwards, and over. This is practically the equivalent of the ordinary wrist throw and will give you possession of the pistol. Turning half left, attack with foot or knee to the fork, open-hand chin-jab, butt with the top of the head, or do anything calculated to knock your opponent out. Each movement has been described separately but, in practice, the movements should be performed so quickly they appear to be almost one.]

b. From the rear.

[Method A. Hands up, well above your head and wide apart. Make up your mind which way you will turn. If to the left, look over your left shoulder, to make sure that it is the pistol that is touching your back. At the same time, as you look over your shoulder, turn your right foot inward. When you are ready to move, turn right around to the left, at the same time bringing your left arm down in a circular sweep over your opponent's pistol arm, continuing the sweeping motion until your opponent's arm is locked firmly under your left armpit. Simultaneously with your turning round, your right hand comes into position for a chin-jab or punch to the jaw and your right knee comes up to your opponent's fork. Finish the matter by turning smartly to your right, reinforcing the movement with your right hand on the elbow of the arm that is still locked under your left armpit. This movement, if continued, will bring him across you and in to position for either a smash to his face with your right knee or a side-of-the-hand blow, with your right hand on the back of his neck. All done with lightning speed.

Method B. Hands up, well above your head and wide apart. Look over your right shoulder, turning your left foot inward as you do so. Turn right around, to your right, locking your opponent's pistol arm, as described above, but under your right armpit. Meanwhile, your left arm is coming around for a side-of-the-hand blow across his throat or face. You are also in position to use your knee. Finish as described above, by turning to your right, etc.]

c. Disarming a man found holding up someone else. [If holding the pistol in his right hand, smash down with your left hand on his forearm at the elbow joint, simultaneously seizing the pistol from underneath with your right hand. Turn rapidly to your left until you are face-to-face with him, pressing the pistol upwards toward him and finally to the left. Use your knee and butt with your hand.]

Students should become proficient in all these five methods of disarming.

5. <u>Searching a prisoner, if you are armed with pistol or rifle.</u>

Kill him first. If that is inconvenient, make him lie face to the ground and search him after you have knocked him out with your boot, your rifle, or the side or butt of your pistol.

6. <u>Taking a prisoner away, if you are armed with a pistol or rifle.</u>

Get someone else to cut the prisoner's belt or braces, or make the prisoner do so himself. March him away, one of his hands above his head, the other holding up his trousers.

7. Securing a prisoner, for some time.

Show the conventional method, using 15 feet of cord and any effective knot, [i.e., place him face down on the ground, tie his hands behind his back, lead the cord around his throat and back to his wrists, around both ankles, and back to his wrists again]. This is excellent, but don't forget to take the cord with you. Also show what can be done with less cord or with belt, braces, or shoe laces. Show how to simply gag him. [Almost anything will do to stuff in his mouth—turf, cloth, a forage cap, etc. For something to tie over his mouth, strips can be torn from his clothing.] All these operations are easiest done if you knock him out first.

[It is useful if instructors will let the students, once or twice during the course, go through the whole process of tying up and gagging, having handy some cord and strips of cloth for the purpose. It is not enough for students to be told how to do it; they must do it themselves.]

8. Arm-break, stressing its simplicity and possible value in crowd fighting.

9. Bent-arm hold.

Using the full length of the forearm to parry, instead of trying to grip your opponent's wrist. Explain that this hold is for use if you have been too slow to get in an effective blow the instant your opponent's arm is raised to strike. Having shown the hold, discard it in favor of the parry and a simultaneous attack, e.g., the with chin-jab and knee. [Why waste time? You have got to parry, in any case, so parry with one forearm and attack simultaneously with the knee and disengaged hand (chin-jab or punch to the jaw).]

[10. Defense against a downward or sideways blow.

Presuming that you are utterly unable to get hold of any kind of weapon, no matter how crude, employ one of the following methods:

a. Side-step and attack.

b. Parry with the opposite forearm and attack.

Students should know, at this stage of the training, how best to attack.]

Note. For the first one or two lessons, students can be in PT kit. Subsequently, it is generally better if they are in full marching order, but without weapons. A German steel helmet, if available, is useful in practicing, for example, the sentry holds.

Various. This is a selection of holds, blows, attacks, etc., which will be known, of course, to every instructor and which should be learned by every student who wishes to qualify as an instructor. In general, they are unnecessary and most of them can be regarded as peacetime measures. Students should be warned that, against a trained adversary, it would be difficult, risky, or impossible to employ some of these holds, etc. It is extremely unwise, incidentally, to assume that one's adversaries in this war will be untrained men. In showing a hold or attack, also show the appropriate release or counter.

1. All the following, good as some of them are, are open to objection that while attempting to apply them, you make yourself very vulnerable to attack. Also, if you are in a position to apply them, you are equally in a position for a killing attack on your adversary:

Handcuff hold.
Handcuff hold for smaller opponent.
Wrist and neck attack.
Police or "come-along" hold.
Flying mare and variations.
Hip throw.
Wrist throw.
Japanese ankle throw.
Cross-buttock.

A further list is given below, each item with comment or objection:

2. Thumb and elbow hold. Show how escape can often be effected. The hold itself may be regarded primarily as a peacetime measure for use by police against untrained opponents.

3. Hand hold. Dangerous because it gives your opponent an opportunity for a crotch throw that can finish you.

4. Arm and neck hold. Effective when got but extremely difficult to get with a quick opponent.

5. Japanese strangle. When the hand is in the correct position, i.e., almost on top of the head, it is impossible to prevent the hand from being dragged away. If the hand is held lower down, where it cannot be seized and dragged away, then there is no leverage and the hold becomes ineffective. It can only be relied on if secured and taken to its conclusion with

extreme speed. [A steel helmet would probably make it very difficult to secure the hold.]

6. Rock-crusher. Good, but only if delivered in exactly the right spot and if there is no equipment in the way. [Why not use one of the other methods of attack?]

7. Grapevine. Useless as a means of keeping a man prisoner. It takes two men to apply it, and if the prisoner does not escape (and some men can), he will die before very long. If you wish to kill him, do so, but don't torture him. If you wish to keep him prisoner, tie him up.

8. Matchbox blow. Excellent, if you happen to have a matchbox at the critical moment. Why not use the side-of-the-hand blow?

9. Baton and spring cosh. Show how to use them and how to avoid them. Both are open to the objection common to all forms of attack with a raised arm. With the spring cosh, if you miss your blow with the extreme end of the weapon, you probably hit your opponent quite harmlessly with the spring. Is either weapon worth the trouble of carrying [it] about? You have got a knife, or should have.

10. Bayonet parries. Show them, by all means. They are all more effective if you, unarmed, are extremely quick and your opponent, armed with rifle and bayonet, is rather slow. [Show particularly the one which consists of parrying the rifle or bayonet away to your left, using your right hand and making a simultaneous half turn, stepping in immediately to your opponent's left-hand side and attacking at speed with hand, foot, or both. Disregard the rifle once you are past the bayonet point.]

11. Safety razor blade, or blades, in cap peak. Show it so that students may know what to expect but do not recommend them to use it.

Note. Holds designed to keep a man captive or to take him away as prisoner. When showing these, it is well to point out that the man who attempts to use them for any purpose other than as a means of finishing off an opponent should realize that he is running a considerable risk that is only justified if he has first crippled his opponent or if he possesses a marked and obvious superiority in physique and knowledge.

NOTES FOR INSTRUCTORS
[Document 3; 1942]

Arisaig Area.

1. Careful preparation of the next day's work is essential. Instructors must have a clear idea beforehand of the lessons to be taught and the methods to be used.

2. The necessary stores, maps, weapons, etc., must be ready before the commencement of the period, and this should be arranged preferably by the NCO whose subject is being taken.

3. Squads must be small and should not exceed seven.

4. Try to arouse the student's interest and then maintain it.

5. Be brief and to the point, but avoid the tendency to sacrifice accuracy and clarity for speed.

6. Endeavor to work up a spirit of competition.

7. When asking questions, do not ask individuals. Address the question to the whole class, and after they have had time to think it over, then ask for an answer from an individual. In this connection suppress tactfully the eager ones who will blurt out the answer.

8. Ensure that the students are all comfortable and that they can see the weapons, etc., used for the demonstrations.

9. All stores, weapons, maps, etc., will be returned clean and in good condition to their appropriate stores immediately after they are no longer needed.

10. Before setting out on a scheme, ensure that the necessary administrative work has been done and checked, e.g., student's clothing, rations, meal on return, transport arrangements, setting of sentries, preparing of objectives, etc.

Remember. Enthusiasm, like yawning, is most infectious, and an enthusiastic instructor will have his class on their toes from the beginning.

ENDNOTES

1. In 1990, the U.S. Marine Corps reprinted Fairbairn and Syke's shooting manual.
2. See listed references by Biddle, Fairbairn, Greene, and Yeaton.
3. British War Office, GS(R), *The Art of Guerilla Warfare* (San Francisco, CA: Interservice, 1981), 8.
4. George Brown, ed., *Commando Gallantry Awards of World War II* (Bournemouth: Bournemouth Press, 1991), 10.
5. F. Spencer Chapman, *The Jungle is Neutral* (New York: W.W. Norton & Company, 1949), 10-11.
6. Jack Evans as told to Ernest Dudley, *The Face of Death* (New York: William Morrow & Company, 1958), 22.
7. See S.P. MacKenzie, *The Home Guard* (Columbia: University of South Carolina, 1995); David Lamp, *The Last Ditch* (United Kingdom: Cassell, 1968); James Ladd, *Commandos and Rangers of World War II* (New York: St. Martin's, 1978); and Peter Young, "The Army Commandos," *The Commando Memorial* (Aldershot: Gale & Polden, 1978).
8. William Pilkington, personal communication, 6 September 1997.
9. William E. Fairbairn, "Anti-Invasion Measures of Defence Tactics," July 1940.
10. Michael Calvert, *Fighting Mad* (New York: Bantam Books, 1990), 43-52. See M.R.D. Foot, *SOE: The Special Operations Executive, 1940-1946* (University Publications of America, 1986); and David Stafford, *Britain and European Resistance, 1940-1945* (Toronto & Buffalo: University of Toronto, 1980). Stafford presents a different view than Foot with some documents, particularly Chapter One, "The Origins of SOE, 1939-1940."
11. Troy, *Wild Bill and Intrepid*, 89; R.H. Barry, "Resistance: Aid From the Outside," *History of the Second World War*, 2(6) 1967, 608.
12. R.H. Barry, "Resistance: Aid From the Outside," *History of the Second World War*, 2(6) 1967, 608. For excellent specific examples, see Pierre Lorain, *Clandestine Operations* (New York: Macmillan, 1983) and the motion picture *Now It Can Be Told* (UK: RAF Film Production Unit, 1946).
13. OSS, *The Overseas Targets: War Report of the OSS*, Volume II (New York: Walker, 1976), 183-185. See also William J. Morgan, *The OSS and I* (New York: Modern Literary Editions, 1957), passim.
14. Calvert, *Fighting Mad*, 100-101; SOE Mission 104 included officer in charge Lieutenant Colonel Mawhood, demolitions instructor Michael Calvert, fieldcraft instructor F. Spencer Chapman, small arms instructor Peter Stafford, and a signals noncommissioned officer.
15. William E. Fairbairn, "Close Combat," n.d., 1 (GHCA).
16. William E. Fairbairn, "Gutter Fighting, The Fighting Knife," 11 February 1944, b (NARA).
17. Fairbairn, "Gutter Fighting, The Fighting Knife," b.
18. Military Intelligence Service, *British Commandos* (Washington, DC: War Department, 9 August 1942), 14; Kelly Yeaton, Samuel S. Yeaton, Rex Applegate, *The First Commando Knives* (Williamstown, NJ: Phillips, 1996), passim; Alan W. Locken, *Collector's Guide to the Fairbairn-Sykes Fighting Knife* (Winnipeg, MB: Author, 1995), passim.

    The knife was purpose built, simple, and relatively cheap (about $3), since commandos were given it to keep. According to Locken, the design was completed in November 1940 and made by Wilkinson in January 1941. The first model had a steel blade drop-forged then hand ground and fitted, with a square ricasso and tang, 2- or 3-inch S-guard hilt, and a checkered brass handle as shown in *Get Tough!* It had a polished blade with nickel-plated hilt and handle.

    The second model was produced in 1941-1942 by various manufacturers. The drop-forged blade was ground all the way to the tang and cut back to fit a flat two-inch hilt. Bright and blackened finishes were used. It was simplified from the first model but exhibited quality construction and balance.

    A third model came out in 1943 and was to become the most copied fighting knife in the modern world. The blade was machine stamped and ground, the handle was from die-cast zinc alloy, ribbed and shortened slightly, and was not as easy to manipulate as earlier patterns. Manufacturers simplified construction using cheaper materials.
19. Fairbairn, "Gutter Fighting, The Fighting Knife."
20. Paul Balor, *Manual of the Mercenary Soldier* (New York: Dell Books, 1988), 78-79.
21. William Pilkington, personal communications, 22 July 1997.
22. Michael Crichton, *The Great Train Robbery* (New York: Alfred Knopf, 1975), 132-133.
23. Joseph E. Persico, *Piercing the Reich* (New York: Viking Press, 1979), 246.
24. George Langelaan quoted in Foot, *SOE*, 64-65.
25. Brigitte Friang, *Parachutes and Petticoats* (London: Jarrolds Publishers, 1958), 14-15.
26. Eric A. Sykes, personal communication, 1 March 1943.
27. Yeaton, et al., *Commando Knives*, 106-107.

28. Rex Applegate, Introduction, *Shooting to Live*, by William E. Fairbairn and Eric A. Sykes (Boulder, CO: Paladin Press, 1987), xiv.
29. See War Office, *Small Arms Training, Volume I, Number 11, Pistol*, (London: HMSO, 1942); *Unarmed Combat*, (London: HMSO, 1941).
30. Bernard Fergusson, *Beyond the Chindwin* (London: Collins, 1945), 32.
31. Calvert, *Fighting Mad*, 100-101.
32. William Pilkington, personal communication, 6 September 1997.
33. M.R.D. Foot, "OSS and SOE: An Equal Partnership?," *The Secrets War*, by George C. Chalou, ed., (Washington, DC: NARA, 1992), 295-300.
34. H. Montgomery Hyde, *Room 3603* (New York: Farrar, Straus, and Company, 1963), 227-228.
35. William Stevenson, *A Man Called Intrepid* (New York: Ballantine Books, 1976), 209-210. See Hyde and Stafford's comments on this source.
36. Michael Elliot-Bateman, ed., *The Fourth Dimension of Warfare* (Manchester: Manchester University Press, 1970), 107.
37. Colin Gubbins, "SOE and Regular and Irregular War," in Elliott-Bateman, *Fourth Dimension*, 107; Hyde, *Room 3603*, 228. A summary of the British and American relationship in setting up the OSS is an appendix by Sir William Stephenson, head of British Security Coordination, in Hyde, *Intelligence Agent*, 247-261.

CHAPTER THREE

FROM COORDINATOR OF INFORMATION TO OFFICE OF STRATEGIC SERVICES

HELP FROM SHANGRI-LA

Two distinct types of training were developed by the COI: one to prepare agents for espionage, ultimately under conditions prevailing in neutral or occupied countries, and the other to provide personnel for various forms of sabotage and to organize guerrillas. Both were set up with help and advice from the British, including instructors and materiel that were parceled out accordingly, with Brooker to Special Activities/Bruce's (SA/B) Rural Training Unit-11, while Fairbairn and Woolley went to Special Activities/Goodfellow's (SA/G) Training Unit.[1]

In fact, SA/G's Lt. Col. Garland Williams had established four schools by April 1942 that followed the SOE pattern: Areas A, B, C, and D. These 10 weeks of special operations training were regulated into phases at these locations, with firearms and close combat as part of the first weeks of instruction.

At Area B were two weeks for assessment with elementary tutelage in physical conditioning, close combat, firearms, demolitions, and fieldcraft. This was followed by two weeks of basic school, which improved on previous instruction and added sabotage and raids. An additional two weeks covered underground and guerrilla techniques.

At Area A near Quantico, Virginia, two weeks of advanced instruction were given that dealt with cover, organization, resistance, and subversion in foreign countries, with another week for parachute training in jumping and receiving supplies by air. A one-week maritime operations school was given at

OSS Area B, Catoctin Mountain Park, Maryland, as it looks today. The cabins were used to house personnel in the isolated and heavily wooded training area. Area B is the location of the present presidential retreat known as Camp David. (Melson collection.)

Applegate (rear, fourth from left) and Fairbairn (front, second from right) at Area B in 1942. This photograph of the American and Allied training staff was taken by Cdr. John Ford's OSS Field Photographic Unit.

Hardware available for sabotage and subversion. A wartime static display exhibits OSS Research and Development Branch products. Silenced firearms are at the center rear and personal weapons at lower right: smatchet, fighting knife, and spring cosh. (Photo courtesy of John W. Brunner.)

Fairbairn brought to Area B his brand of gunfighting, which was adapted and developed by the schools branch of the OSS. He is at the left demonstrating the low firing position to a batch of attentive students. (Photo courtesy of NARA.)

Hand-to-gland methods of close combat were presented as formalized dirty fighting at Area B. Fairbairn demonstrates how to break a hold using Applegate as a training dummy. (Photo courtesy of NARA.)

Applegate demonstrates disarming methods on his own "victim" to ensure that even unarmed personnel could overcome an armed opponent in extreme circumstances. This and the previous two photographs were from director John Ford's official film OSS Training Group. (Photo courtesy of NARA.)

Area D on the Potomac River at Smith Point, Maryland, to cover landing by boat or submarine. A three-week industrial sabotage course was available for some at Area A and separate communications preparation was undertaken at Area C.[2] The OSS official history notes that the training program was never static, "but was constantly being developed, refined and improved."[3]

Area B was located in what is the Catoctin Mountain Park near Thurmont, Maryland. During the war, about 9,000 acres were controlled by the federal and state government and had been restored by the depression-era Civilian Conservation Corps (CCC) to an eastern hardwood climax forest. The CCC built a central park road, trails, and three semipermanent camps useful for classrooms and housing year-round. The uneven, forested terrain was isolated but still convenient by rail and road to Baltimore and Washington, DC. The park service helped to get the facilities going and Applegate reported it suitable for the paramilitary course of instruction envisioned by the special operations arm of COI.

An army commanding officer had overall responsibility for Area B and a chief instructor ran the training. A staff of about 40 put through groups of around 15 students on an almost individual basis. The close-combat training curriculum developed at Area B formed the basis for later OSS programs: the approach was oriented specifically to combat rather

Inseparable from Area B was the presidential retreat known as Shangri-La. In residence from 1942 to 1944 were national and international dignitaries, including Prime Minister Churchill shown here fishing with President Roosevelt. (Photo courtesy of FDR Library.)

than self-defense, with techniques presented to kill or disable efficiently and speedily. The result was an offense-oriented form of close combat using firearms, edged or clubbed weapons, and hand-to-hand fighting (in that order) to create or avoid a confrontation. It reflected the diversity the OSS needed to prepare individual undercover operatives and uniformed military special forces, including the occasional familiarization by staff or headquarters personnel with a taste of "tradecraft."[4]

From 1942 through 1945, Fairbairn was in the United States "on loan"[5] from BSC with the Schools and Training Branch of the OSS that succeeded the COI on 13 June 1942—a joint civil-military organization that evolved into the post-war CIA. Applegate recalls meeting Fairbairn at Area B.

> I first met Fairbairn when he came down from Canada and we soon sized each other up, a lieutenant and a captain. That evening there was a demonstration for assembled COI dignitaries in the mess hall building. He called me out on the stage and said, "I want you to attack me." The thought of a 28-year-old 230-pounder going against a 57-year-old 160-pounder was incredulous to me. He continued, "Just like you were going to kill me." I let out a roar and went for him and wound up sailing through the air into the audience. I had been in some barroom brawls and held my own, but it got my attention.[6]

From that point on, the two worked to put a practical close-combat program together for the civilian and military trainees of this unique service. The Americans easily accepted the newly promoted Fairbairn and knew him as "Major Foul-Blow" for his "honest dislike for anything that smacked of decency in fighting."[7]

Jerry Sage served as an Area B instructor and knew Applegate from his Ft. Lewis days. Sage recalled the Camp Two facilities as being log cabins, a swimming pool, and a large athletic field. A loudspeaker seemed to be going all the time, blaring out music and messages. The initial staff recruited and sent to Area B all had a part in turning the trainees, mostly civilians, "from jailbirds to college professors," into functional soldiers. Once this was accomplished, Sage wrote:

> We turned our emphasis on the tough training—physical conditioning, explosives work, hand-to-hand combat and knife fighting. Ultimately, every man learned how to survive as an agent behind enemy lines, how to collect intelligence on the enemy and get it through to our side, and how to sabotage the enemy's efforts against us.[8]

Other Ft. Lewis alumni were Christian (Chris) Rumberg and Joseph Collart. With college football and gymnastics backgrounds, they designed and built a "trainasium" obstacle course from logs, ropes, and embarkation nets to overcome fear of heights and to "gain confidence in their own agility and sense of balance."[9]

Fairbairn gave both basic and instructor classes assisted by Applegate and the others. Used was the "Fairbairn Method," based upon his experience in Shanghai and the United Kingdom and *Get Tough*. The system was briefed as standard for British forces and growing in use in the American forces. The training was intended to overcome the "natural repugnance" to killing and to teach dirty fighting to accomplish the mission. Instructors exhorted, "Let yourself go—kill, deceive, sneak, thief, be unfair, be blood thirsty, unleash your hatred . . . let your weapon be an appendage, a part of you." Hit with the automatic, butt stroke, cosh, newspaper, broken bottle—"know all the tricks."[10] Applegate and other instructors were told that this:

> . . . is a time for the quick or the dead. If you're not the former, you'll end up being the latter. To attack is to have the battle half won, while to stay on the defensive is fatal. But attack alone is not enough, the element of surprise must go hand and hand with it. To surprise is to strike fear into the enemy's heart and fear results in confusion.[11]

Fairbairn felt that you could not "consider yourself an expert until you can carry out every movement instinctively and automatically." This resulted from daily practice, slowly at first then with increased speed. In the words of one lesson plan, "Killing at close quarters demands aggression and extreme concentration."[12] He asserted that "all these methods have been specially selected as being the most practical for students about to take up work with this organization. Every one of these methods is practical, and has, on many occasions, stood the acid test of actual combat in the present war."[13]

Another early close-combat instructor was Thomas L. Curtis, who was trained with nine others in 1942 and later went on to fight in occupied Greece. They worked for five days on close-combat, pistol, revolver, carbine, and Thompson submachine gun techniques to qualify as instructors and assistant instructors. Fairbairn recommended that they be used in pairs until experienced, and that they be provided a kit to include a fighting knife, cosh, and a Colt .38 "Banker's Special" revolver. A two-day refresher course was also desired within three months.[14] They learned "the rough-stuff department, how to kill with a small rock, a pencil, or even a folded newspaper" and were "able to fire any small weapon at the end of the course." Of this grounding, an instructor commented:

> I spent so much time with the close-combat course of the legendary Major Fairbairn, a mild-mannered Britisher who had been Shanghai's chief of police and cleaned up the world's toughest

> gangster colony, that I was afraid to spar with my friends for fear of maiming them.[15]

The Area B basic close-combat course of instruction included armed close combat (using the knife, smatchet, and miscellaneous weapons), unarmed close combat (using blows, releases, and holds), and battle shooting (using the pistol, submachine gun, and carbine). Follow-on advanced OSS courses expanded the use of weapons, close combat, obstacle courses, and calisthenics. For special operations and operational group personnel, training included the full range of military small arms, of which the rifle, submachine gun, pistol, and mortar were "most useful," and a full range of Allied and Axis firearms.[16] Actual time spent in the instruction of close combat varied and was based more on relative economy than anything else. In the United States, the average for OSS basic was about seven hours for close combat and 14 hours for point/instinct shooting of the pistol, submachine gun, and carbine.[17]

Each subject was supervised by a training instructor of commissioned rank and such assistants as were required to present the lesson effectively. Emphasis was placed on the necessity for thorough preparation of each lesson on the part of the instructors and their assistants. This included a thorough knowledge of available references and a detailed plan for conducting a class. Applegate and his fellow instructors learned the technique of military instruction that consisted of lecture, demonstration, application, and simulation-evaluation. They were taught to "tell them what you are going to teach them, teach them, tell them what they were taught," and to *keep it simple, stupid* (KISS). Other lessons were learned as they gained experience with students.

Classroom work took up varying amounts of time, but an average, most probably, of 30 minutes. The rest of the period involved practical exercises with individual coaching of students by instructors. Every effort was made to stimulate interest, enthusiasm, and understanding on the part of the students. Lectures were short and to the point, covering only such necessary instruction that could not be presented by practical demonstration. Constant effort was made to introduce realism in training as conditions permitted, and to develop initiative and leadership through the conduct of practical problems in minor tactical situations. Instilled was the idea of stalking, stealth, kill or be killed—deceit, subterfuge, dirty fighting, everything that is underhanded and vile in order to accomplish the mission. For this the OSS weapons and close-combat trainers' presentations developed along the following lines:

OSS CLOSE-COMBAT INSTRUCTION

1. Create enthusiasm by competition wherever possible.
   a. Time obstacle course in gym.
   b. Team work while firing on the ranges.
   c. Sell the type of firing taught.
2. Develop leadership in men.
   a. Call on students now and again to briefly discuss some phase of the instruction.
   b. Let men coach each other on the firing line.
   c. Let men take charge of giving the firing commands when qualified.
3. Foster cooperation by allowing men to assist each other and the instructor wherever possible.
4. Keep records of all scores on ranges whenever possible. Note aptitudes and abilities of students as they work.
5. Observe each student individually and as a worker in the group. Learn the names of all the men as quickly as possible.
6. Develop proper attitude. Impress the following on the men throughout the course.
   a. Never be unarmed—substitute weapons.
   b. Be alert—observe.
   c. Get tough—now is the opportunity.
   d. Get to know your weapon.
7. Fire the lightest gun in its class first in order to overcome gun shyness.
8. Keep ranges policed—after firing have men pick up brass and do not allow old targets and pieces of rubbish to lie around.
9. Never point a weapon into the air when not being fired. Point at ground.
10. Zero all weapons from the prone position.

Presentation

1. Call roll. Introduce self to class. Begin memorizing the names of the men. After each class (throughout the course) jot down outstanding characteristics of each man in accordance with the final report form.

Orient the Class

1. This department is responsible for:
   a. Weapons
   b. Close combat
   c. Physical training
   d. Obstacle course
2. Weapons course—students will be able to fire any small arm by the end of the course.
3. Calisthenics—reveille at 0615—assembly at 0630. Fifteen minutes of exercise prepares men for a hearty breakfast. Men will get out of the course what they put into it. Get into condition.
4. Obstacle course—four times in three weeks. Do not undertake what you cannot accomplish first time around. Build up to it.
5. Close combat—men are shown the best blows, throws, holds found in jujutsu, judo, etc. A body-building course was given for those who really like to develop their bodies scientifically.

Important Points to Remember

1. Blank ammunition is dangerous. Illustrate with story of accident.
2. Care of equipment. Keep it clean and use it with care. Treat your weapon better than you treat yourself. It is your best friend.
3. Observe. Observe men and objects around you. Keep your eyes open and your mouth closed. Students are being observed every minute by every instructor and students should observe their fellow students.
4. Instructors wish to assist the students in any way possible. If any problem comes up, do not hesitate to ask for advice. Ask questions.

The Fairbairn Method of Firing

Remember that the majority of killings in close combat firing take place within 4 yards—stairways, alleys, roof tops, etc. One and a quarter seconds ends most gun fights. One magazine is sufficient.

Conclusion

1. Mission will govern type of equipment carried. The most useful weapons will be either some type of rifle, carbine, submachine gun, pistol, or mortar.
2. Know your weapon. Learn its "feel." Obtain dexterity.
3. Never be unarmed. Remember substitute weapons. Demonstrate the correct method of hitting with the pistol (not with the butt).
4. Get into condition while you are here. This will be your last opportunity in this country. Exercise while on the way on the boat.[18]

Applegate and his companions at Area B were not alone in recalling the impact made with this program. For those that went through it, anecdotal accounts abound even half a century later. For some students this was in passing, for others it was the basis for further qualifications, and for a very few it was a matter of life and death.

Max Corvo, who later led OSS activities in Italy, recalled having to travel incognito by train from Union Station in Washington, DC, to Baltimore, and then on to Sparks, Maryland. Picked up there by an army truck, he was deposited at Area B with some 15 other students, all using assumed names. The training course "had been set up with the help of SOE officers, among them was a Major Fairbairn." Fairbairn and staff were "reputed to be the foremost experts in hand-to-hand combat." The experience was mostly physical to test endurance and courage: long marches, the obstacle course, compass courses, demolitions, and small arms "were the order of the day."[19]

Ib Melchior remembered his Area B experience, arriving at Area B with his fellow trainees in the middle of the night. The next morning, their chief instructor walked them by a mock cemetery with a waiting open grave. The army captain lined the students up in front of a mound of earth and pulled his .45 Colt from its holster, briefing them on its nomenclature. He concluded by stating that there was a big difference between being at the grip or muzzle end: "He suddenly opened fire. The live bullets whizzed through our group, zinging across our ears to slam with a deadened thud into the dirt mound behind us."[20] Some students froze, others dropped, and some took off running with a resultant decrease in class size.

Gunfighting was recalled under the title of the "Quick or the Dead" and described as "snap" shooting. This was for close-in (5 to 40 yards) fighting where "the first shot with a reasonable amount of accuracy" counted, producing what Fairbairn called "good, fast, plain shots."[21] In working with Fairbairn's shooting technique, "the important thing is to ensure that the weapon is fired the instant the shooter faces the target. Particularly in training, there is a tendency for the individual to attempt to correct his aim after facing in the direction of the target and before pressing the trigger."[22] Hopefully this would not happen with a target that shot back! It was given as a complement to, not a replacement for, aimed fire, and it was practical for jungle or street fighting. Point shooting was learned with the pistol, submachine gun, rifle, and shotgun: students were taught how to fire anything "from a pistol to a BAR at close quarters, by aiming with the body."[23]

Aaron Bank, an OSS officer who fought in France, wrote that he practiced instinctive firing: "a method by which you aimed a handgun with your body rather than with your eyes on the sights—a specialty for night combat."[24] John K. Singlaub described his training experience at Area B, before going on to Europe and the Far East. He felt Fairbairn stressed instinctive weapons use and at the range commented, "You chaps have been trained to aim a weapon. That's all proper and good on the range. The only problem is that your average Hun or Nip rarely stands still with a bull's-eye on his nose." He proceeded to demonstrate this by spinning on three pop-up targets and letting go six shots in quick groups of two. The man and the pistol had become a single weapon: "in a crouch, his right arm cocked with the pistol just forward of his hips, his left arm also cocked as a balancing boom, and his knees slightly bent."[25]

During this shooting phase, cryptographer Dr. John Brunner, who served in China, and a partner were practicing with the pistol. Brunner got the hang of instinctive shooting, but his partner did not and got Fairbairn's attention without much success for a full afternoon.[26] Applegate noted on the *OSS Training Group* film that Fairbairn "did not correct the bent shooting arm if it was close enough—Fairbairn was best with the small group and not the mass production that took place elsewhere."[27]

Sage wrote: "The ultimate test of our instinctive shooting ability was Fairbairn's specially designed house of horrors, which we faced at the end of the course."[28] Melchior described this in more detail.

> It was like a huge carnival shooting arcade—except you were in the games. Targets would fly up, each to receive two rounds. Sacks filled with sawdust would lurch at you. Trap doors, loose floorboards, closed doors, and blind corners would bedevil you. At one point, as you kicked open a closed door, a man would face you, his gun at the ready. But—better not fire. It was your own reflection in a full-length mirror.[29]

This shooting house was made unpredictable and was keyed up by rumors to make the participant uncertain as to how he would react and to demonstrate his chances of survival in "a real life-threatening situation."[30] Aline Griffith wrote about an incident in Spain that underscored the usefulness of this preparation.

> I turned and without hesitating crouched and shot at the shadowy form rushing at me, 15 feet away. I must have missed, because he lunged for me and grabbed my throat. I shoved the gun at him, not knowing or caring where I was aiming this time. I was sinking into unconsciousness when I heard the sound of a shot. My own![31]

Neither armed nor unarmed combat was neglected at Area B. Three Fairbairn and Sykes "toys"[32]—the fighting knife, smatchet, and cosh—were imported from the United Kingdom, and the OSS Procurement and Supply Branch contracted local manufacturers to produce similar items. The specific techniques of employment taught at Area B made these weapons useful tools in the clandestine war against Germany and Japan, rather than the curios they are today.

The OSS claimed that the Fairbairn knife fighting system was the most widely known technique among American combatants: "It embodies various slashing operations, with aim directed at vulnerable arteries at side of neck, lower center of stomach, heart, below shoulder blades, above wrist, or at armjoint."[33] Applegate related that "Fairbairn went around in British battle dress with his knife on his leg and would pull it out to make his points."[34] Applegate wrote:

> He frequently, and unexpectedly, would draw the knife from its concealed location and place the point at either the throat or stomach of various OSS personnel whom he was meeting for the first time. Needless to say, he got their attention and left more than a lingering impression. One night at the Army and Navy Club in Washington after a number of scotch and sodas, he stuck himself in the thigh while sheathing the knife. What the British call a "bloody mess" ensued.[35]

Sage said that "Fairbairn was about twice my age. He stood 5 feet 10 inches and weighed 170 pounds. In martial arts training he always picked a bigger man to attack him to prove that the smaller man could win if he learned Fairbairn's techniques." These included how to fall, roll, "and come up on the offense." They became skilled at come-along grips and blows that "could dislocate a man's shoulder or break and arm or leg." Sage recalled the most "lethal silent kill as the Japanese strangle or sentry kill." Instruction encompassed the stiletto-type knife, machete, and smatchet, "a hybrid between a machete and an axe."[36]

Singlaub continued: "Our principal weapons and unarmed combat instructor was British Maj. William Ewart Fairbairn of the legendary Fairbairn-Sykes team." In particular, once you felt "the old man's" callused grip on your throat when he tripped you savagely to the ground during a feeble knife lunge, "you realized you had a lot to learn." Fairbairn's fighting creed was simple: a well-trained man had nothing to fear from close combat, rather, "if this man was properly armed, all nearby adversaries had everything to fear."

Bank added:

> We had good martial arts instructors and we specialized in knife fighting and throwing, a silent form of killing. [This consisted of the] OSS mixture of catch-as-catch-can and judo. We had been taught how to approach a sentry from the rear, snap an arm around his neck in a choke hold, and thrust a stiletto of Fairbairn design between his upper ribs while bending him backward. Major Fairbairn was also highly visible in the sunken gardens where we practiced silent killing with the use of the knife or garrote. Knife fighting was included.[37]

Brunner told how Fairbairn gave his training group close-combat classes and recalled two things. First, Fairbairn's advice for a knife fight was to shoot or run. He selected the largest private present, gave him a knife, and said, "Kill me." The private was not going to attack a lieutenant colonel until Fairbairn kicked him in the "nuts" and provoked the desired attack to be countered.[38]

Richard Helms, a later director of the CIA, reminisced that secret intelligence personnel, using only their first names, learned to handle themselves in life or death situations. Fairbairn "taught us the deadly arts. Within 15 seconds I came to realize my private parts were in constant jeopardy." His whole theory was "that gentlemanly combatants tended to end up dead, and he persuaded us that this was the proper attitude in the area of self-defense."[39]

Instructor Roger Hall put this knowledge into practice in a confrontation with counterintelligence agents in an American city.

> It was over seconds after it began. Gordan threw one gorgeous right hand which caught Berg

> square on the nose and sent him sprawling. Connor came at me, grabbing a chair on the way. It was as though Major Fairbairn were standing beside me. "And if they pick up anything, such as furniture, wait until they lift it to strike you, then move in and go for the belly."[40]

Fairbairn recounted to Applegate and others the process that close combat contributed to.

> In initiating training in close-contact fighting, "gutter" methods in offensive and defensive tactics were taught to the personnel assigned for such training. With the building up of a natural confidence in the ability of each man to perform duties of this nature, their training began to show results and their efficiency continuously increased. From reports from overseas theaters, it has been noted that irrespective of whether or not the personnel who have undergone this type of training make use of it in the missions to which they are assigned, the mere fact that they have been so trained and have been imbued with the confidence which comes from it, their efficiency in other types of combat operations, and their morale, is greatly increased.[41]

In early 1942, Catoctin Mountain was selected as a location where President Franklin D. Roosevelt could get away from Washington for his own well-being and private consultations with a select group of leaders. The chief executive preempted the use of nearby CCC Camp Three, soon nicknamed "Shangri-La," an illusory place from a James Hilton novel and the location from which the Doolittle raiders were said to have bombed Tokyo (the bombers were actually off of an aircraft carrier).[42]

Meeting with Special Agent Michael Reilly, Applegate "did the first security survey with the Secret Service when they went up to scope it out for the president's use. Up to this point the U.S. Secret Service had never seen a tree." Camp Three was a rustic hideout of 18 cabins, a shower block, swimming pool, office, craft shop, and baseball field with little hot water and brush growing up to the window sills. The president arrived on 22 April 1942 to approve the site, the first of some 20 visits during the war. Work on the 143-acre compound began by personnel from the park service and military.

One of the initial overnight visitors with the president was the OSS' Donovan. Applegate remembered: "FDR expressed an interest in the activities adjacent to Shangri-La and a demonstration was put on by the staff, including Fairbairn and myself. He sat in a wheelchair behind a sheet of bullet-proof glass obtained in Pittsburgh." The *OSS Training Group* film was one result of this and was originally planned for Roosevelt's private use. Applegate later believed this was part of the "power struggle" between Donovan over intelligence scope and territory "with J. Edgar Hoover of the FBI, the U.S. Army, and Navy intelligence."[43]

Both Fairbairn and Applegate appeared in the *OSS Training Group* film directed by John Ford. Another instructor recalled that acting for Commander Ford "was terrific fun. He filmed us swinging from trees, throttling a sentry and blowing up huts and enemy material. All trainees wore masks to disguise their identities."[44] Some OSS recruits caused presidential security concerns as some were expatriates and radicals "of questionable backgrounds by Secret Service standards." For six weeks Applegate served as an additional bodyguard when Roosevelt was in residence, because ostensibly Applegate knew who the students were. He had been first noticed by the president because of his size and the butt of a .45 FitzGerald Colt revolver sticking out of his back pocket. In a subsequent interview, Applegate's politics were noted by the president to ascertain if partisan politics in Oregon extended to his duties.[45]

Applegate recalled with particular pleasure witnessing Roosevelt and Churchill at close quarters during the latter's June 1942 visit to Washington, DC, at a time when support of North Africa, the need for a second front, and atomic weapons were on the agenda.[46] Observing a particularly heated discussion, Applegate asked the chief of the White House Secret Service detail what to do. "You let 'em fight it out themselves," was the agent's reply.[47]

The Secret Service supervised the installation of fences, floodlights, sentry boxes, and gates to be manned by U.S. Marines when the president was visiting. When the president was not present, the Marine guards served as voluntary or involuntary aggressors for the various field problems run by the OSS training staff. As Applegate put it, the Marines were busy in the Pacific and "most of those I worked with were officers who had been wounded and just returned to duty. There were very young and inexperienced enlisted Marines at Shangri-La. They were used for drills and simulated assassination exercises when FDR was not around that eventually resulted in the use of dogs and chain-link fences by the Secret Service."[48]

Hall wrote that the Area B designation stood for "By God, it's a long way from nowhere." He felt the presence of Shangri-La and the Marines made for more realistic field problems day and night because "they seldom fired more than twice before yelling 'Halt!'" They were "joking" and when the OSS staff got close enough to remind them that they were all on the same side, the answer was, not to worry, "We fired over your heads."[49]

BRITISH SPECIAL OFFICERS COURSE

In the summer of 1942, joint training with the British was given to Americans by the SOE and Secret Intelligence Service (SIS). Before being deployed to occupied areas, the British provided operational support for them as well. Later in 1942, overseas training was conducted for OSS personnel in North Africa and the Far East.

Applegate remarked: "I stayed at Area B while Fairbairn made the rounds of the other OSS training areas that were opened up. I put through the first two or three classes before being sent to the United Kingdom."[50] Others who accompanied Applegate included Maj. Ainsworth Blogg, Capt. Christian Rumberg, Lt. Charles (Charlie) Parkin, Jr., and Lt. Arden Dow, all fellow officers from Ft. Lewis now serving with the OSS. As Applegate remembered it: "I was one of the first two sent from the OSS, with Charlie Parkin, to Fort William and the commando and SOE facilities. Fairbairn and Sykes had started there with the home guard then moved on to the SOE and special training centers."[51]

This exchange tour took Applegate on the rounds of SOE, SIS, and Combined Operations training and facilities with other Americans eager to learn what the British had been doing for the last few years. Applegate went first to Fort William and training at Arisaig, Lochailort, and Achnacarry. Here experts were assembled from all over the commonwealth and allied sources. Applegate found that "the most skilled and famous of these men were also engaged in training special intelligence personnel for operations in enemy occupied territory. Schools were operated under tight security conditions all over the English countryside and in northern Scotland, where country estates of the nobility were used as training centers."[52]

By 1942, the Special Training Center at Lochailort was named the Advanced Infantry Assault School to train officers and non-commissioned officers in two to four weeks to become instructors in close combat, demolitions, and fieldcraft. Of 95 hours of training for close-combat instructors, 14 were allocated for unarmed combat (the last six hours were for evaluation as instructors), eight for sabre fencing (because the footwork, balance, mental reactions, and moves were suitable for knife fighting), and an hour of bayonet fighting. Two hours were allocated to the .45 automatic pistol and submachine gun for handling and mechanical training, and two hours apiece for firing, with an hour "hip firing" the rifle. Applegate's British instructors found him a weapons training instructor "who is an expert in his job, but takes little interest in anything else."[53]

Americans found that former members of the Shanghai police "gave the instruction in unarmed combat, the use of the fighting knife, and self-defense." It seemed close combat had become a form of physical training, requiring the use of all muscles and improved bodily coordination. "Many hours were devoted to unarmed combat, involving jujutsu, wrestling, and general brawling tactics. This training improved the individual's self-confidence and developed a keen desire to fight." The fighting knife developed at the school was carried by all members in a special pocket on the uniform trousers. A very difficult pistol course was arranged, which "required the students to fire from various angles at unexpected targets." Instruction was given to fire the submachine gun with single shots while it was set for full automatic to conserve ammunition, while being available for automatic fire.[54]

This was where Applegate met William Pilkington, who was a protégé of Gubbin's and one of Fairbairn and Sykes' first students from STC days (a former sergeant commissioned as a close-combat instructor). The early days were remembered when close combat appeared to be the only option available and Pilkington's "highly lethal staff, stick, and baton system" was advocated by Fairbairn.[55] Pilkington also suggested the use of improvised weapons from clipboards to docker's bailing hooks. "Some of my ideas were blacklisted by our War Office as un-British, crude, unthinkable to consider such brutal methods." Pilkington felt that Fairbairn:

> . . . was very much in charge. Captain Sykes had equal authority and great ability. He was the finest rifle shot I have ever seen, as well as being very good with the .45 Colt 1911 automatic pistol. Both officers were very skilled in unarmed combat also. I did teach a few hundred people the killing arts, and I am grateful for the training I experienced with Fairbairn and Sykes, they were really masters of their craft.[56]

Applegate became one of the first members of the OSS to go through the SOE "pipeline," along with other American participants who described the training as it appeared mid-war. Operatives attended a number of separate courses arranged to fulfill individual assignments, and this served as a series of filters to their fitness for duty. Evaluated were leadership, physical fitness, language skills, and knowledge of small arms.

The paramilitary Group A schools in Scotland were at Arisaig, where students were assigned to a medium-size manor house. For about two weeks they underwent ubiquitous psychological tests and practiced physical training, marksmanship, and self-defense taught by "former members of the Shanghai Police." Food and drink were plentiful, considering the times. Instructors who had served with French, Yugoslav, and Greek guerrillas ran the training program. Course material was guerrilla operations and tactics, enemy weapons, and demolitions. Map and compass courses were given day and night for up to 20 miles over wind-swept moors and ridges. "The grounds were overgrown, and nearby hedges and outbuildings had been adapted as weapons and demolition training areas," recalled Applegate.[57] Here Applegate met Maj. Eric A. Sykes and his perspective on close combat.

> My first meeting with Sykes was short on conversation. Sykes outfitted me with a brand new Sten gun and a Webley revolver and had me run through a combat course in the basement of a medieval Scottish castle. He had installed a comprehensive combat range complete with targets in German uniforms that fired blanks at the trainee, booby traps, and other devices. Apparently, I did well enough because a close relationship with Sykes followed.[58]

Training included nine and a half hours with the .22 and .32 Colt automatics and 10 hours with the Thompson and Sten submachine guns, which were fired indoors and outdoors at static and moving silhouette targets on lit and unlit ranges. Other practices included outdoor stalk-and-attack training in the open. The object of the last was to develop speed in the attack. The targets were in view from a distance and the attacker dealt first with one target from the prone position at about 30 yards. He then closed quickly to 20 yards and dealt with the next target from the long-range standing position. Finally, he moved quickly to close quarters and fired at the remaining three targets from the orthodox attacking position. The sequence of the above varied according to the ground, but as much variety as possible was included and the close-quarter targets were wide apart to give maximum right and left swing. Speed and aggression were stressed as much as possible.[59]

At the time, Applegate described this in his own words.

> A spectator watching one of the famous British assault courses in

which live charges, live grenades, and live rounds of ammunition are fired around the man participating in the course, asks himself if he would actually be able to take such a course. From the spectator's viewpoint, it looks very spectacular and the element of danger thrown in by live ammunition's striking close to his feet, charges bursting around him, and all the other real battle effects is very impressive.

This same spectator, once he enters upon such a course, is so intent on firing his own weapon, throwing his own grenades, and reaching his objective that he does not heed or notice the various charges bursting around him while he is going through the course. In a large way this explains a soldier's reaction in combat. He is so intent on his own job or mission that, after the initial effect, he is not bothered and does not think about what is going on around him.[60]

By 1942, all commando replacements were trained at the Commando Basic Training Center at Achnacarry, where Applegate went next (along with some 25,000 others who passed through six or more weeks of instruction). According to the director of Combined Operations, the Earl Mountbatten of Burma, this was just one of the three stages necessary for this type of fighting: first "the whole idea has to be thought up and thought out, secondly the right type of volunteers have to be obtained, and thirdly highly specialized training must be arranged."[61]

The school's commander, Lt. Col. Charles E. Vaughn, had his own views toward the arrival of the Americans: "Gentlemen, we must make it hard for them, very hard indeed." One instructor interpreted this to mean, "No juke boxes, no Coca-Cola, no blondes, truly a sad situation for a red-blooded son of the United States to find himself in."[62] On arrival, Applegate and the other Americans were immediately put through a speed march on the "commando dark mile," a 5-mile course to be completed in less than 50 minutes. For Applegate, this was accomplished carrying a Thompson submachine gun with the drum magazine bouncing into the base of his spine along with his web gear.

A British instructor, James Dunning, wrote that: "From the moment they arrived at Achnacarry, the commando trainees set about learning the tactics and techniques of war and devising new ones. No detail was too insignificant, no plan too bold, yet foolhardiness was frowned upon." The emphasis was on night training, operating in bad weather, crossing rough terrain, and the assault landing. Live ammunition was used to provide "battle conditions."[63] Despite this level of intensity, one American observer felt the average U.S. Marine was "better trained than even the specialized commando in the elementary subjects as well as fieldcraft."[64]

Lt. Col. William O. Darby and the first U.S. Army Rangers went through the depot at the same time, and he wrote about it in detail.

The commando's first item in his trilogy of training was physical conditioning, developed by speed marching and execution of runs over increasingly difficult obstacles courses. Next we learned to use the portable weapons, including rifle, machine gun, bayonet, and mortar, which were needed by hit-and-run raiders. There were no specialists in the camp. All men had to meet intentionally high standards in employing these weapons. The third requirement was battle preparedness—the expert ability to execute small tactical problems.[65]

The depot provided its own combat firing and close-combat instruction from a former London policeman at odds with what had been taught earlier by Fairbairn and Sykes, but to the liking of the depot commander. Darby described it: "There was boxing and close-in fighting. There was no particular emphasis on jujutsu, though the men were given a few good usable holds that each could be expected to remember and utilize when needed."[66] Douglas Fairbanks, Jr., a naval officer on the Combined Operations staff, benefited from some of this by "learning to shoot with a tommygun (when necessary) [and] to stab and wrestle" in preparation for amphibious deception operations with allied forces. He observed one use of the fighting knife not prescribed in training at Achnacarry: at meal time food was "pulled apart with regulation commando daggers or pocket knives."[67]

After basic commando training by the army, the Royal Navy carried out needed amphibious preparations at Argyll. This included actual combat seasoning; OSS men went along with a U.S. Ranger detachment on Operation Jubilee at Dieppe—a major hit-and-run assault on a heavily defended area of France. For Applegate, this was with 1 Commando in July 1942. As he saw it: "At the time the commandos were the only action the British had going and I bootlegged my way onto an operation. Most were relatively small time."[68]

An official historian wrote that many of these undertakings were "unreported in the press, in which a landing was made by perhaps a dozen men, a sentry or two killed, a prisoner or two brought back. On these occasions the damage done and the loss of life inflicted were as negligible as the information obtained."[69]

With a raid in the offing, designated units carried out special training. They pored over maps, sand tables, air photographs, and drawings of an unnamed area until there was confidence in routes and locations. Exercises were carried out over similar terrain as rehearsal after rehearsal took place. Many a proposed raid fizzled out, so the guess that action was imminent was discouraged. Much depended on the weather, tides, moon phases, and available shipping; as a commando recalled "we wavered between hope and despair."[70]

A move by truck to a "concentration" camp followed at the port of embarkation where transport was boarded. Boat drill was practiced to file from compartments to landing craft without a hitch. Ammunition was issued and grenades primed, in the midst of which everyone would be summoned to the mess deck for last minute instruction once at sea.[71]

Applegate's most vivid recollections were that the weapons used for training were the same taken on operations. He told writer Pete Dickey about his excursion, recalling that instead of the usual Thompson submachine gun, his group was issued Sten guns with only a brief familiarization the night they crossed the English Channel. Applegate then related how they crept through enemy outposts en route to the target: "Communications was all but absent, as there was only starlight and even low voices were taboo."[72] As the party crossed a low ditch, the first commando jumped into it and his tightly gripped Sten went off, firing not just one round but a full magazine! This "was the most amusing part" of the action, said Applegate. Having just made a costly withdrawal under fire the previous month, the raid leader decided not to carry on against alerted Germans and extracted the unit. Talking about this operation afterwards, Applegate's instructors from Scotland recalled that he "does not know the meaning of fear."[73]

There was parachute training at STS-51 at Altringham, near Ringway Royal Air Force Base. This training was low key, averaged three days, and was based on the fact that it was a mode of transportation and not an act of heroism. Qualification jumps were required using a "Joe" hole, the first from a captive balloon at 800 feet. The Americans witnessed two French women and a 65-year-old man ("Jills" and a "Joe") make their final jumps wearing tennis shoes! Major Blogg and Applegate inspected the school and demonstrated unarmed combat, while Parkin undertook the course with flying colors. Applegate missed jumping because of a previous leg injury and later made a parachute jump from a tethered balloon into a pond.[74]

More exposure ensued from "Broadway to Baker Street," which included sophisticated close combat added to that from Scotland. Group B finishing schools ensued near Beaulieu Abbey, along with specialist courses. Students rotated around London and the Midlands for work on intelligence activities—cover stories, agents, and networks for security, communications, sabotage, subversion, counterespionage, and escape and evasion. One dealt with the forming of resistance and guerrilla organization, and another specialized on raids and advanced demolition techniques that focused on the sabotage of railroads. Charges were placed on tracks, locomotives, tenders, and freight cars, or were disguised as coal and mixed with the real thing. The London street-fighting school in the bombed-out East End was used. Raids were performed on buildings up to three stories high, using teams to control the stairwells and halls while other cleared rooms.[75]

Sykes continued the close-combat training. For example, he called the operatives out on "a really dark, moonless night" for sentry-elimination drill. A student approached the uniformed dummy in the approved manner and drove his knife into a knapsack that had been absent on previous exercises. "You had to determine before the attack whether a knapsack was being worn. A two-man elimination team was the safest and quietest, since one man effected the assault while the other grasped the sentry's rifle before it dropped."[76] Another American recalled:

> Our close-combat instructor was now Maj. Bill Sykes, Fairbairn's partner. Sykes' method of instinctive firing training involved creeping around a blacked-out cellar and shooting at moving targets illuminated by the muzzle flash of our first shot. Once we could consistently hit the targets with absolutely no hesitation, Major Sykes pronounced us "improving."[77]

Of this, Applegate wrote: "I spent many informative hours talking weapons with Sykes. He also did me some personal favors and arranged for some special training that, as far as I am concerned, is still classified to this day."[78] This involved methods perfected by, among others, Lt. Col. L.H. Grant-Taylor, who's credo was *surprise is critical to winning a gunfight*. He was documented as having walked into a room containing six Luftwaffe pilots and their female companions and shooting them all (the pilots, that is) within 20 seconds of opening the door. These techniques were practiced with and without ammunition until the shooter had perfect confidence in his weapons, allowing him to react quickly and concentrate on the kill. With this, Sykes came into his own, counseling to clear a room from the bottom of the door frame to provide a difficult target—don't stand in the doorway and never enter a room without a full magazine.[79]

The British reported that Lieutenant Applegate was "an excellent instructor and, although there is no humbug about him, he is rather inclined to showmanship. It is possible that he would get better results from students of his own nationality, who would understand and respect him."[80] This experience was significant to Applegate's development, particularly in the relationship with Fairbairn and Sykes, and the formulating of his own subsequent close-combat efforts.

Applegate's recollections of this are worthwhile to consider at length.

> Fairbairn and Sykes were good. This was a new world for me, particularly in Great Britain. I tried to get the best out of everything. Sykes was very capable and knew more about firearms, but was brainier and laid back and he looked like a clergyman. Fairbairn was the bare-hand and knife man, and an extrovert.[81]
>
> Both Fairbairn and Sykes were close to retirement from the Shanghai police when they were called to active duty and given the rank of captain. They were declared too old for combat duty and were assigned to training operations. Stories abound, however, of how they tried to circumvent their noncombat classification in favor of front-line service. By 1942 they were not the best of friends. Sykes had prevented Fairbairn from going on a raid because of his age and Fairbairn got publicity from his books and the choice assignment to North America.[82] In all the time I was with Fairbairn in the states, he never mentioned Sykes by name. When I returned to the U.S. we did not discuss my training with Sykes.[83]
>
> Looking back over the years to my associations with these famous fighting men, I have several lasting impressions. Fairbairn was the most flamboyant and aggressive of the two. He had a more basic interest in unarmed combat and knife-fighting techniques. Sykes, a more reserved individual, had more expertise in firearms. Both excelled in all

> phases of close combat with and without weapons.[84]

Applegate returned to the United States via neutral Portugal in August 1942. He continued to refine instruction at Area B, based upon what he had learned from Fairbairn and Sykes and wartime developments to get "the biggest bang for the buck." Applegate carried on his relationship with both men, one in regards to close combat, particularly knife work, and the other in relation to firearms, particularly with silencers.[85]

Following Donovan's charge "to learn all that there is to learn" about close combat, Applegate had conclusions of his own. One central belief was that the best way to dispatch an opponent was the quickest and easiest way for the fighter himself.[86] These solutions to close combat were based on a genuine appreciation of the enemy.

> The German doesn't like close combat: he relies on his weapons and fire power, and his whole psychology is built around the thought of throwing out a lot of lead to keep from closing with the enemy. Since boyhood the Jap has been trained in jujutsu. But the important thing to remember is that the Jap thinks he is the best fighting man in the world. This makes him aggressive and dangerous.[87]

By the end of 1942, Area B proved to be too far from Washington for efficient administration, and a further drawback was its being across the road from the president's "summer" White House; when Roosevelt was present, activities had to be considerably curtailed. The "rough" CCC camps and summer recreational facilities were "far from adequate through the rainy and cold seasons." Area B's functions were taken over by other OSS schools, after establishing the model for similar programs.[88]

Applegate was due for a change: "I put through two or three classes upon returning from the United Kingdom. As a regular army officer I was advised to get out of the OSS and get a real job" with the U.S. Army.[89] This change saw Fairbairn remain with the OSS, Sykes serving in the United Kingdom with the SOE and SIS, and Applegate establishing himself with the MID; all were expanding in their own way upon the close-combat techniques developed first in Shanghai.

The following document reflects the basic instruction in firearms and close combat developed from 1942 to 1943 at Area B, and what Applegate took from Fairbairn and Sykes to his next assignment.[90]

NOTES FOR INSTRUCTORS IN CLOSE COMBAT
[Document 4; n.d.]

The course consists of two sections—Armed and Unarmed

Armed: 1. Knife fighting with the fighting knife, jackknife, and penknife.

2. Fighting with the Fairsword [Smatchet].

3. Fighting with the cosh or stick.

4. Fighting with a newspaper, etc.

Unarmed: 1. Attacking methods.

2. Releasing methods.

3. Overpowering methods.

All these methods have been specially selected as being the most practical for students about to take up work with this organization. Every one of these methods is practical, and has, on many occasions, stood the acid test of actual combat in the present war.

Although there are hundreds of different methods (some good, some poor, some absolutely useless), it is inadvisable in a short course to attempt to teach more than what is shown above. It would only result in the student being confused and not gaining any practical knowledge. Instructors will therefore confine themselves to instruction in the above methods only, and, as far as possible, present them in the order shown.

In this connection, it is inadvisable to spend too much time demonstrating disarming a man holding you up with a pistol. Instructors must be very firm with students on this point; otherwise they will spend all their time learning a purely theoretical method (in war, your opponent does not "hold you up"—he "shoots on sight"). Further, this method of disarming is now so widely known by practically everyone, that it is now of no value.

Instructors must, during the course, be 100 percent on the job. They must sell and inspire confidence at all times, treating each student as a personal friend whom they wish to help. Every student is a volunteer with a man-sized job to perform, and it is the instructor's responsibility that everything possible that will assist him in making a success of his mission is given cheerfully and willingly. The following was the highest praise ever received by any instructor; at the conclusion of a successful commando raid on Norway a telegram was received [from Lord Lovat]: "Dear ___, it works!"

The repugnance to killing with a knife as compared to killing with a pistol is one of the most difficult problems facing the instructor. The difference is mainly psychological and a good instructor will be able to break it down and also to immediately build up confidence in the knife, not only as a method of attack, but as the best, and we might say the only method of defense. That is, of course, where firearms are not used.

Instructors must at all times when instructing knife fighting keep in mind the following WARNING:

A. It is impossible to state what will be the reaction of any two men to the flashing of a knife in close proximity to any part of their body.

B. Students under instruction will always be in single line 6 to 8 feet apart.

C. Instructors will never approach a student with a naked knife without first informing him of the intention and telling him not to move.

D. Do not permit any student to demonstrate on another the Knife Searching Method. Instructors, when demonstrating this method, will do so with the <u>pointed end of a pencil</u>. Explain to the student the content of above, paragraph A., and the likelihood of a student falling backward on the knife and being permanently injured.

CLOSE COMBAT, PART A, ARMED—"Shoot first, that is why you are issued a pistol!"

Sequence of instruction, notes for instructors:

<u>Method No. 1 The Fighting Knife.</u>

A. THE GRIP—Grip with the index finger and thumb, edges to right-left. Knuckles underneath.

B. LEFT PARRY—Parry toward the left hand side with a circular sweep of the wrist and arm.

C. LEFT-RIGHT PARRY—The reverse hand—knuckles [. . .] and parry to the right-hand side.

D. LEFT-RIGHT PARRY WITH THRUST—Thrust, by lunging with the left foot and extending the arm.

NOTE: Start all students with the boxing stance. Later it is immaterial which foot is advanced. Have frequent rest periods and roll the knives. Students should change knives. This brings out the point that no two knives have the same feel.

<u>Method No. 2 Attacking a Kicking Opponent.</u>

LEFT-RIGHT SLASH WITH THRUST TO NECK—Immediately contact is made on opponent's foot or leg—the fingers are clasped around the handle and the cut is made by a circular withdrawing motion.

Method No. 3 Shadow Fighting.

A. ON UNEVEN GROUND—In a field or up and down a slope or hill or around a house. Students should use every type of cut or thrust—keeping on the move all the time.

B. UP AND DOWN STAIRS, etc.

NOTE: During Shadow Fighting only one student will be permitted to practice at a time. Others will be formed up behind.

NOTE: Inform students no unarmed opponent can disarm a skilled or semiskilled knife fighter. Stress this point very strongly.

Method No. 4 The Jackknife and the Penknife.

A. UNDERHAND GRIP—Cutting edge on top—attack by an upward motion from the testicles up to the chin. Show method of hooking the opponent's neck with the left and pulling forward into the above, "rip."

B. OVERHAND GRIP—Demonstrate how the Finns settle their disputes—holding each other's hair with left hand.

NOTE: Don't belittle this method of fighting, the Finns are good at it.

With the penknife, withdrawing slashes should be on the side of the neck or across the nose and on face. Demonstrate how a small knife can be concealed in the inside seam of the pants above the boot.

Method No. 5 The Fairsword. [Smatchet]

A. THE CHOPPING GRIP—Firm grip with all fingers and thumb. Cutting edges up and down.

B. ATTACKING BLOWS—1. Wrist or arm; 2. Thrust to stomach; 3. Sabre Cut from high left to right side of neck; 4. Cutlass Cut from high right to left side of neck; 5. Smash up under the chin with the pommel; 6. Smash down into the face with the pommel.

NOTE: Blows 3. and 4. will sever the carotid arteries which are only approximately 1 and 1/2 inches below the surface.

C. KNIFE GRIP—All methods taught for the fighting knife can be accomplished (with practice) equally as well with the Fairsword. Hold with firm grip of index finger and thumb—cutting edges parallel with the ground.

NOTE: The Fairsword, on account of its size, is a uniform weapon scientifically constructed to permit the balance being adjusted so that it will have a perfect balance, no matter what size of hand. It can be used for cutting one's way through the jungle, chopping firewood, or any other domestic work usually performed with an axe. There is a small knife, pick, bottle opener, screw driver and a stone for sharpening attached to the scabbard.

Method No. 6 The Cosh, The Stick.

A. COSH GRIP—Adjust the wrist cord to permit the head of the cosh to shoot into center of the palm of the hand. Carry the cosh concealed up the sleeve of the coat with the "nob" flush and between the index finger and thumb.

B. ATTACKING BLOW—First extend the spring and measure the length, plus the arm length, then depress the cosh and conceal up the sleeve. Contact must be made with the weighted end of the cosh on the side of your opponent's head anywhere between the cheek bone and the ear, upwards, a space of approximately 24 inches—a very easy matter with a little practice.

C. THE STICK—Break a stick—1 inch by about 20 inches—from a tree, thus demonstrating there is no justification for anyone in the country to be without any weapon. 1. The grip; 2. Blow to stomach with side of end; 3. Blow up under chin with end; 4. Blow down into the eyes or face; 5. Side blow to head; 6. Two-handed blow up to chin.

Method No. 7 The Newspaper Attack.

A. HOW TO FOLD—Take a double sheet (comic section) down to approximately 6 inches by 2 inches; then fold tightly cross-ways forming a sharp point at one end.

B. HOW USED—1. The grip—point forward between thumb and index finger; 2. Strike with point into stomach immediately upwards under the jaw about 1 or 2 inches back of the chin with all the strength you can put behind the blow.

C. ALTERNATIVE METHOD—For attack when sitting on the right of your opponent. Take three or four double sheets of newspaper and fold down about 8 inches—then roll into a cylinder shape. 1. Hold with one [hand]; 2. Strike with a circular outward and inward swing of your right arm at left-hand side of opponent's face. Anywhere between the chin and temple will knock him unconscious.

CLOSE COMBAT, PART B, UNARMED—"For those who have been foolish and caught unarmed."

Sequence of instruction, notes for instructors:

Method No. 1 Edge-of-the-Hand Blows.

POINTS OF ATTACK—Forearm bone, wrist, side of neck, one inch below Adam's apple, back of neck, etc.

NOTE: Fingers and thumb straight out. Strike with little-finger edge of hand.

<u>Method No. 2 Chin Jab.</u>

POINT OF ATTACK—1. Up under the chin with the heel of the hand—fingers extended to reach the eyes.

NOTE: This is a close-in blow and must be given with all possible force in an upward direction.

<u>Method No. 3 Tiger's Claw.</u>

[POINT OF ATTACK]—1. Into the eyes and face with a piston-rod forward jab. Fingers bent like a tiger's claws. Weight of body to be behind the blow with no telegraphing of the intention to strike. This is the most effective hand blow ever worked out. It has the advantage of 3 inches of additional length as compared to a straight left and can be achieved with lightning-like speed. It is a complete answer to any attempt at a frontal attack and permits one to deal effectively with one's opponent before he is really dangerous.

<u>Method No. 4 Kicking.</u>

METHOD OF KICKING—Turn right foot sideways to the left and kick forward. Make contact on the shin bone with the outside edge of the sole of the boot. Follow through by transferring the weight of your body from the left to right foot and smash down with your boot onto the small bones of opponent's foot.

SPECIAL NOTE: 1. Any student who thoroughly masters the three hand blows and the method of kicking (no matter what his strength or build may be) will be able to effectively deal with any unarmed opponent; 2. All these methods are <u>attacking</u>, <u>not defensive</u>, and should be applied in conjunction with each other.

<u>Method No. 5 Arm Jerk.</u>

1. Seize the opponent's arm at the wrist with both hands (right above left); 2. Lift the arm up about 6 inches and jerk it downwards.

WARNING: This must in practice be applied very mildly. If applied with force, concussion or delay concussion will result. When applied on opponent's right arm the force of the jerk will be felt on the left side of the head.

<u>Method No. 6 Smacking the Ears.</u>

Cup your hands and simultaneously smack both of your opponent's ears.

NOTE: Have students practice on themselves, only applying it very mildly. If applied with force, one or both ear drums will burst.

CLOSE COMBAT, PART C, UNARMED—"For those who were caught, releasing methods."

Sequence of instruction, notes for instructors:

Method No. 1 Release from a Wrist Hold.

A. ONE-HAND HOLD—Bend wrist in a circular motion out and over the thumb.

B. TWO-HANDED HOLD (thumb on top)—Catch hold of your seized wrist and jerk upwards toward you (against opponent's thumb).

C. TWO-HANDED HOLD (thumb underneath)—Reach under and catch your hand and jerk downwards toward you. If pulled toward your right will break opponent's thumb.

Method No. 2 Release from a Hair Hold.

FROM BEHIND—1. Bend backwards—Seize his right hand with firm grip; 2. Turn sharply on your left foot towards your left-hand side and sharply pull his right hand down to the ground (nevermind if you pull your own hair out); 3. Simultaneously bring your right foot as far to the rear as possible; 4. Retaining a vice-like grip on his wrist, kick him in the face.

Method No. 3 Release from Bear Hug.

A. FROM BEHIND (arms pinned)—Sink suddenly downwards by bending your knees and grab his testicles with your right hand. Grip hard and jerk downwards.

B. FROM IN FRONT—Go after his ears with your teeth simultaneously bending and grabbing his testicles with your right hand.

NOTE: These are the only methods that will permit you to break the bear hug of a really strong man and they must be applied quickly, before you are thrown to the ground, and with force.

Method No. 4 Release from Come-Along Hold.

HOW HELD—1. Your opponent has locked your arm by seizing your wrist and placing his arm over and under yours. A rather painful hold; 2. Do not resist. Get out of step with him and suddenly jab the side of his leg with your nearest leg.

CAUTION: This must be done suddenly by throwing your body [weight] onto your attacking leg. You will most likely receive a very bad sprain of all the muscles of your seized arm. Your opponent cannot break your arm, but his leg will be broken for certain. This method should only be demonstrated by the instructor.

Method No. 5 Release from Collar and Arm Lock.

HOW HELD—1. Your opponent has bent your left arm up your back and has a strong grip on your coat and collar at the back of your neck; 2. Set the muscle of your neck to take the strain off your throat and with the greatest possible speed twist completely around toward your right-hand side, simultaneously striking an edge-of-hand blow at your opponent's head with your right hand.

SPECIAL NOTE: Students should be warned that the release from the come-along hold and collar and arm lock may be very painful, but they can be assured that it will be much more painful to their opponent.

Method No. 6 A Reliable Come-Along Thumb Hold.

SPECIAL NOTE: Students will be warned that there is a great difference between arresting a man in normal times, and taking a man prisoner in wartime. In war, a man will take five times more punishment and exert five times more strength to prevent capture. It is impossible to secure an effective hold on a prisoner until you have at least partially knocked him out.

HOW APPLIED—1. Seize his left arm at the wrist with both hands and apply the Arm Jerk; 2. Right palm up, insert your right thumb between the thumb and forefinger of his left hand, maintaining a very firm grip; 3. Seize his left elbow with your left hand, palm down, and simultaneously force his forearm up his chest and pull the elbow over the right forearm; 4. Force his arm into your body with your right arm and seize his fingers with your left hand and apply pressure by pulling them downwards toward your left-hand side.

WARNING: This is the most painful hold known and from which there is no escape and all students should master it thoroughly. Even Japanese judo experts admit there is no escape from this hold, especially if one's opponent has first been brought to a submissive frame of mind.

NOTES FOR INSTRUCTORS ON PISTOL SHOOTING

The course consists of practical war methods of shooting with the one-hand-gun, with which any man of average intelligence can be taught to draw, load, fire, and hit his opponent within a second.

This is not a new-fangled idea, but a proven method which has been in use in the Far East since 1919 against some of the most desperate criminals in the world—men who were, in most cases, known killers who preferred always to shoot it out rather than be captured and finish up in front of a firing squad. Records kept over a period of 12 1/2 years in which only one-hand-guns were used show the following: 666 shooting affrays—260 criminals killed, 193 wounded; 42 police killed, 100 wounded.

All affrays were on the run—up and down stairways, over roofs, down cobbled alleyways, or in very crowded streets. Ninety percent of the shooting was in the dark and the majority of the hits were made within four yards. Some of the police were killed by being shot in the back at a matter of inches distance only.

Students should be informed that the average shooting with the one-hand-gun is over, so far as they will be concerned, in a very few seconds. There will be no time to reload. If their first shot takes longer than a third of a second to fire, they will not be the one to tell the newspapers about it. It is literally a matter of the quick and the dead—so they can take their choice.

Every movement (manner of holding the pistol, position of the feet, etc.), right from the start, is a part of the training to gain speed in firing and the instructor will insist upon students carrying them out to the letter.

NOTE: Where necessary, demonstrate how any other method does not permit one to fire with speed.

WARNING: 1. Instructors will not permit the pistol to be handled, loaded or unloaded in any other manner than that laid down in these instructions; 2. Instructors will always stand on the left-hand side and slightly forward of the student when he is on the firing point at the initial instruction. During the rapid load and mystery shoots, etc., he will be close-up and immediately behind him.

PISTOL SHOOTING

Sequence of instruction, notes for instructors:

First Day.

[Dry Firing]

A. [SAFETY PRECAUTIONS] PROVING THE PISTOL—This must always be carried out in the following manner. 1. Remove the magazine; 2. Work the slide; 3. Release the trigger. CAUTION: Whenever a pistol is taken in hand it must be immediately proven in the above manner.

B. LOADING—1. Prove the pistol as above; 2. [Pistol rolled to the right.] Insert the magazine and TEST; 3. [Pistol rolled to the left.] Hold the slide with the left hand, arm bent at elbow, push forward with the right hand in a 45-degree angle toward the ground; 4. When the right arm is fully extended, release the hold of the slide with the left hand and permit it to drive the cartridge from the magazine into the breech.

NOTE: This method is much easier than pulling the slide with the left hand, and must be insisted on.

C. UNLOADING—1. [Pistol rolled to the right.] Remove the magazine; 2. [Pistol rolled to the left.] Work the slide; 3. Release the trigger.

CAUTION: No pistol is unloaded until the above has been carried out in the order shown.

D. THE READY POSITION—1. Gripping the pistol in the right hand (finger inside the trigger guard), arm straight, rigid and across the body, bend the hand slightly to the right to bring the pistol in line with the vertical centerline of the body; 2. Raise the pistol arm (still rigid with a 20-pound grip) straight up from the shoulder until the pistol covers the aiming mark; 3. Wait a pause—pull the trigger, then count to three slowly and lower the arm slowly to the Ready Position.

NOTE: The foregoing should take approximately 20-30 minutes. There must be frequent rest periods, during which the instructor should give the reasons for the 20-pound grip and finger on the trigger, etc.

Firing Practice.

[Initial practice at 9 feet; 2 types of target, recruit and standard. The recruit target is 8 feet by 8 feet with a man-size silhouette in the center; the standard target is also 8 feet by 8 feet with a 1 x 1 1/2 inch solid black rectangle in the center, and concentric 4 x 8, 8 x 12,and 12 x 16 inch rectangles.]

[THE STANCE]—1. Standing erect—feet apart—square to the target—LOAD; 2. Raise the pistol—make a pause—fire one shot—count to three—lower the pistol. Then repeat and fire another shot; 3. Bring both feet together—advance the left foot, as if walking—go in to a crouch leaning forward onto the toes—pistol at an angle of 45 degrees to the ground; 4. Upon the command FIRE—raise the arm quickly—make a slight pause and fire two very rapid shots—holding the pistol on the target until ordered to lower the arm slowly; 5. Repeat and after firing the remaining two shots—UNLOAD.

NOTE: Instructors will correct the grip of the pistol if both these shots are not within 3 inches of the aiming mark. After all students have fired the first practice—call them to the target and explain what would be the results of shooting at 25 yards and 50 yards.

2d Practice.

1. Load and take up the Crouch Position; 2. On command FIRE—raise the arm and fire two rapid shots; 3. Repeat; 4. Repeat and UNLOAD.

NOTE: Special attention will be paid to keeping the pistol on the target after firing each burst of two shots. Also to the results of each burst. Watch for a dipped wrist—poking—and holding the arm too low.

3d Practice.

1. On the command FIRE—load, go into a Crouch Position and fire two rapid shots; 2. Lower the arm—turn half right to the target. Upon command FIRE—turn the body quickly (square with the target) raise the arm and fire two shots; 3. Lower the arm—turn half left to the target. Upon command FIRE—turn the body—raise the arm and fire two shots. Then UNLOAD.

4th Practice.

Repeat No. 3 Practice.

Second Day.

1st Practice.

Repeat No. 4 Practice.

2d Practice.

Repeat in the dark.

3d Practice.

Repeat No. 4 Practice with lights.

4th Practice.

Where necessary—repeat—others fire two handed, etc.

NOTES FOR INSTRUCTORS .45 THOMPSON SUBMACHINE GUN

The .45 Thompson submachine gun is a close-combat weapon (within 20 yards), and, although the cartridge will kill at 1000/1200 yards, it is only guaranteed for accuracy up to 50 yards. Its main selling point is its volume of fire (morale builder of the operator).

Instructors will insist at all times that safety precautions laid down are very strictly carried and in the order shown.

A few minutes of instruction is sufficient to instruct any student to fire single shots with the firing action set on full automatic. The object of this instruction is to demonstrate that the firing action should never be set on semiautomatic; also that bursts of fire should never be more than three shots.

Always have a 4-inch hole in the "stomach" of the target in the initial firing practice (all shots through the same hole).

The fact that the majority of the shots do go through this hole very quickly convinces the student that he is a master of the weapon.

All firing will be under the direct order of the instructor, except the last practice, which will be fired with the instructor immediately behind the student, where he will be in position to prevent the student from pointing his gun too far to the sides, etc.

THOMPSON SUBMACHINE GUN

Sequence of instruction, notes for instructors:

<u>No. 1 Practice.</u>

[Dry Firing]

A. SAFETY PRECAUTIONS—Prove the weapon by removing the magazine and working the breech. See that the levers are set at the FIRE and AUTO positions.

B. THE STANCE—1. Demonstrate the Carry Position—gun held under the right arm—barrel pointing at a 90-degree angle toward the ground; 2. Insert the magazine and TEST that it is locked into position; 3. Ready Position—Advance the left foot, as if walking—go into a crouch holding the gun parallel to the ground, with a 50-pound grip of the right arm and left hand. Weight of the body forward on the left foot, both heels off the ground; 4. LOAD—Pull back the breech with the left hand; 5. FIRE—in short bursts of one or two (never more than three); 6. UNLOAD—Remove magazine, work the breech, leaving it in the forward (closed) position.

NOTE: The trigger finger should be stiff, not bent, and pressure applied on the right edge of the trigger. If the finger is wrapped around the trigger, it is impossible to release it before four or five shots have been fired. Never, in practice, load magazines with more than 10 rounds.

<u>No. 2 Practice.</u>

[First at a range of 5 yards, then 7 yards, then 10 yards.]

Feet together, magazine inserted, weapon at the Carry Position under the right arm. 1. Upon the command LOAD pull back the breech, advance the left foot and go into the Crouch Position; 2. Fire a short burst of one or two (not more than three) shots; 3. Turn half right, still in the crouch position—On command FIRE turn the body by the quickest possible means and fire a short burst; 4. Turn half left and repeat; 5. Continue until magazine is empty, then UNLOAD.

<u>No. 3 Practice.</u>

Feet together, magazine inserted, weapon at the Carry

Position under the right arm; 1. Upon the command FIRE—load, go into a Crouch Position and fire a short burst; 2. Continue as in No.2 Practice and then UNLOAD.

NOTE: This will be one, continuous movement, with speed.

No. 4 Practice.

Feet together, magazine inserted, weapon at the Carry Position under the right arm; 1. Upon the command FIRE—load and fire a short burst; 2. Permit the students to fire the remaining cartridges in short bursts as in No. 3 Practice without waiting for the command FIRE.

NOTE: When the gun is loaded, it will always be parallel to the ground.

NOTES FOR INSTRUCTORS [M1] .30 CARBINE

With the introduction of the .30 carbine, the days of the submachine gun are numbered and it is not unlikely that the pistol and the revolver will be replaced by this weapon (except in special cases where concealment is essential).

The .30 carbine can be fired under any conditions where it would be possible to fire a pistol or submachine gun and with far greater accuracy. It fires a cartridge with a target range of 300 yards and has a muzzle velocity of 1950 feet per second as compared with:

.45 Automatic Pistol, target range 50 yards—muzzle velocity 820 fps.

.45 Submachine Gun, target range 50 yards—muzzle velocity 940 fps.

NOTE: Instructors will note that although the same cartridge is fired in the pistol and the submachine gun, the muzzle velocity is different. This is accounted for by the increase in the length of the barrel. Also, the submachine gun is fitted with sights up to hundreds (600) of yards. If you want guaranteed accuracy beyond 50 yards, it is advisable that you use a rifle cartridge, not a pistol cartridge.

The .30 carbine is a semiautomatic, reloading weapon. Its balance or feel immediately appeals to everyone. For this reason, instructors will leave the instruction of this weapon to the last day. Where your duties will be in uniform, your selection will be the .30 carbine.

[M1] .30 CARBINE

Sequence of instruction, notes for instructors:

No. 1 Practice.

[Dry Firing]

A. SAFETY PRECAUTIONS—1. Prove the weapon by removing the magazine, work the bolt and release the trigger; 2. Demonstrate the following firing positions—standing, kneeling, lying, sitting, and hip shooting.

B. LOADING—1. Prove the weapon; 2. Insert the magazine and TEST; 3. LOAD.

NOTE: Do not ease the bolt forward—leave go when it is in the rear position and permit it to drive the cartridge from the magazine into the breech.

<u>No. 2 Practice.</u>

[First at a range of 5 yards, then 7 yards, then 10 yards.]

THE STANCE—1. Standing position, at a 2 inch mark on the target—1 shot/1 second—repeat quickly; 2. Hip position, at the hole in the stomach of the target—1 shot/1 second—repeat quickly; 3. Kneeling position, at hole in stomach of the target—1 shot/1 second—repeat quickly; 4. Standing position, 4 rapid aimed shots at 2-inch mark, and then UNLOAD.

<u>No. 3 Practice.</u>

A. Fire deliberate aimed shots at a 2-inch mark, from any position at any distance.

B. If necessary, and time permits, repeat No. 1 and No. 2 Practices.

NOTE: The .30 carbine is a close-combat weapon in which the firing should be snap shooting, with the greatest possible speed (other than No. 2 Practice).

END NOTES

1. OSS, *War Report I*, for SA/B training see 76-77; for SA/G see 80-82.
2. OSS, *War Report I*, 80-82, 322-325. For a very generalized account of SOE and OSS training, see Ian Dear, *Sabotage and Subversion: Stories from the Files of the SOE and OSS* (London: Arms & Armour, 1996).
3. OSS, *War Report I*, 243.
4. McDonald, "The OSS," 24-32.
5. Director OSS, Service Resumé of Lt. Col. William E. Fairbairn, 23 March 1945, passim (NARA).
6. Rex Applegate, personal communication, 30 October 1995.
7. Richard Dunlop, *Behind Japanese Lines* (New York: Rand McNally, 1979), 84.
8. Jerry Sage, *Sage* (n.p.: Miles Standish Press, 1985), 12-14, 20-21.
9. Sage, *Sage*, 21.
10. OSS, "Weapons and Close Combat," 21 December 1943, 2, 26 (NARA).
11. Fairbairn, "Gutter Fighting."
12. Hal Bergman, "The Deacon Deals in Death," *Leatherneck*, June 1944, 50-51; David Stafford, *Camp X* (New York: Dodd Mead, 1986), 97-98.
13. Fairbairn, "Gutter Fighting."
14. William E. Fairbairn memo, 3 December 1943 (NARA).
15. Roger Hall, *You're Stepping on My Cloak and Dagger* (New York: W.W. Norton, 1957), 39.
16. Fairbairn Resumé, 1.
17. Schedules for OSS RTU11, Area A, Area E, Area WA, 1943 to 1945 (NARA).
18. OSS, Instruction Outline, Area A4, August 1943; Weapons and Close Combat, 21 December 1943; Weapons Course, Area A4 (NARA).
19. Max Corvo, *The OSS in Italy* (New York: Praeger, 1990), 14-16.
20. Ib Melchior, *Case by Case* (Novato: Presidio Press, 1993), 8-18.
21. Stafford, *Camp X*, 98.
22. Wallace M. Greene, "The Quick or the Dead," *Marine Corps Gazette*, August 1984, 71.
23. Dunlop, *Behind Japanese Lines*, 84.
24. Bank, *From OSS to Green Berets*, 26.
25. John K. Singlaub with Malcolm McConnell, *Hazardous Duty* (New York: Summit Books, 1991), 31-32.
26. John W. Brunner, personal communication, 30 August 1995.
27. Rex Applegate, personal communication, 30 October 1996.
28. Sage, *Sage*, 14, 19-20.
29. Melchior, *Case by Case*, 8-18.
30. Melchior, *Case by Case*, 8-18.
31. Aline, Countess of Romanones, *The Spy Wore Red* (New York: Random House, 1987), 246.
32. Fairbairn Resumé, 3-4. See John W. Brunner, *OSS Weapons* (Williamstown, NJ: Phillips Publications, 1994). This provides a superb background on OSS weapons and equipment.
33. OSS, *OSS Weapons* (Washington, DC: Research and Development Branch, June 1944), n.p. (NARA).
34. Rex Applegate, personal communication, 30 October 1996.
35. Rex Applegate, Introduction, *Commando Dagger*, by Leroy Thompson (Boulder, CO: Paladin Press, 1985), vii.
36. Sage, *Sage*, 14, 19-20.
37. Bank, *From OSS to Green Berets*, 26-27.
38. John W. Brunner, personal communication, 30 August 1995.
39. John Ranelagh, *The Agency* (London: Weidenfeld and Nicolson, 1986), 66-67.
40. Hall, *You're Stepping on My Cloak and Dagger*, 79, 102.
41. Fairbairn Resumé, 1-2.
42. D. Dale Nelson, *The President is at Camp David* (Syracuse: Syracuse University Press, 1995), 4-21; of interest for what it does not reveal is Edmund H. Starling, *Starling of the White House* (New York: Simon & Shuster, 1946) by the head of the Secret Service presidential detail through 1944.
43. Rex Applegate, personal communication, 12 December 1994, 30 October 1995; Nelson, *The President is at Camp David*, 6-7.
44. Sage, *Sage*, 33.
45. Pete Dickey, "Connoisseur of Close Combat," *American Rifleman*. March 1993, 38-39.
46. Winston S. Churchill, *The Second World War* (New York: Golden Press, 1960), 184-185.
47. William L. Cassidy, *Quick or Dead* (Boulder, CO: Paladin Press, 1978), 99.
48. Rex Applegate, personal communication, 30 October 1995.
49. Hall, *You're Stepping on My Cloak and Dagger*, 34.
50. Rex Applegate, personal communication, 30 October 1996.
51. Rex Applegate, personal communication, 30 October 1996
52. Rex Applegate, Introduction, *Shooting to Live*, by William E. Fairbairn and Eric A. Sykes (Boulder, CO: Paladin Press, 1987), xiii.
53. Rex Applegate, SOE/OSS memo, n.d.

54. Military Intelligence Service, *British Commandos*, 33-39.
55. William E. Fairbairn, "Anti-Invasion Measures of Defence Tactics," July 1940, 2.
56. William Pilkington, personal communication, 6 January 1996; see also Rex Applegate, "E.A. Sykes," *The First Commando Knives*, by Kelly Yeaton, et al. (Williamstown, NJ: Phillips Publications, 1996), 79-81.
57. Singlaub, *Hazardous Duty*, 35.
58. Applegate, Introduction, *Commando Dagger*, viii.
59. Weapon Training Syllabus, n.d.
60. MITCVIII, Hand Gun Offense, 22 January 1943, 11.
61. Donald Gilchrist, *Castle Commando* (London: Oliver & Boyd, 1960), ix and passim.
62. Gilchrist, *Castle Commando*, 74, 77.
63. James Dunning, "The Commandos at Achnacarry," *The Commando Trail* (Fort William and Lochaber, n.p., n.d.).
64. Joel D. Thacker, *The Marine Raiders in World War Two* (Washington, DC: HQMC, 1954), 3.
65. William O. Darby, *Darby's Rangers: We Led the Way* (Novato, CA: Presidio Press, 1980), 30.
66. Darby, *Darby's Rangers*, 30.
67. Douglas Fairbanks, Jr., *A Hell of a War* (New York: St. Martin's Press, 1993), 153, 160.
68. Rex Applegate, personal communication, 30 October 1995.
69. Christopher Buckley, *Norway—The Commandos—Dieppe* (London: HMSO, 1952), 164.
70. Gilchrist, *Castle Commando*, 108.
71. Gilchrist, *Castle Commando*, 108-109.
72. Dickey, "Connoisseur of Close Combat," 40.
73. William Pilkington, personal communication, 15 November 1994.
74. Singlaub, *Hazardous Duty*, 34-42; Bank, *From OSS to Green Berets*, 18-23; Rex Applegate, personal communication, 16 June 1994. Morgan's *OSS and I* is a good account by a training staff member and later operative. See also S.J. Lewis, *Jedburgh Team Operations* (Combat Studies Institute, 1991) and William Casey, *The Secret War Against Hitler* (Washington, DC: Regnery Gateway, 1988).
75. Applegate, "Sykes," *Commando Knives*, 75.
76. Banks, *From OSS to Green Berets*, 27.
77. Singlaub, *Hazardous Duty*, 39.
78. Applegate, "Sykes," *Commando Knives*, 81.
79. Frederic Sondern, "Murder is His Business," *Reader's Digest*, November 1943, 68-71; Cal Tinney, "Hector's Job was Murder," *True Spy Stories*, ed. by Robert Deindorfer (Greenwich, CT: Fawcett Publications, 1961), 66-77.
80. SOE/OSS memo.
81. Rex Applegate, personal communication, 30 October 1996.
82. Rex Applegate, personal communication, 30 October 1996.
83. Applegate, "Sykes," *Commando Knives*, 82-83.
84. Applegate, Introduction, *Commando Dagger*, vii.
85. Eric A. Sykes, personal communication, 1 March 1943, 22 September 1943.
86. Rex Applegate, California Rangemasters Association, Burbank, CA, 21 March 1996; Sage, *Sage*, 36.
87. MITCVIII, Introductory Lecture, May 1945, 1-3.
88. CIA, Information on OSS Schools and Training Sites, 1 February 1951, 1, 3 (NARA).
89. Rex Applegate, personal communication, 30 October 1995.
90. Other copies of this same document are credited to William E. Fairbairn by name; the inclusion of the Fairsword indicates a date of at least 1943; annotated on the text used here was "Course given at B5 by Dan."

CHAPTER FOUR

THE MILITARY INTELLIGENCE DIVISION AND TRAINING CENTER

SECTION EIGHT SOLUTIONS

Problems with location and the establishment of more suitable facilities made OSS Area B a backwater by the winter of 1942. Applegate, as a regular army officer, felt he needed to get back to his own service. With his OSS and SOE background, he was quickly selected by MID to conduct similar instruction at the newly created Military Intelligence Training Center located at Camp Ritchie, near Cascade, Maryland, a former Maryland National Guard post located just over the hill from Area B.

The War Department's MID underwent wartime expansion as Gen. George C. Marshall felt the intelligence service had to improve to meet the demands of modern warfare by concentrating its various resources. Its aim was defined by a wartime member of the OSS and Counterintelligence Corps: "Military intelligence is the gathering of information about the enemy—his plans, strength, his operations—evaluating it, and disseminating it. Counterintelligence is preventing the enemy from gathering such information about you and carrying out clandestine operations against you."[1]

Fort Ritchie's monument to its wartime function as the MITC, which included Applegate's Combat Section. The German antitank gun was one of those used to prepare intelligence personnel to recognize enemy equipment. (Melson collection.)

By June 1942, the U.S. Army took over the facility to form a consolidated intelligence training center. Camp Ritchie covered about 638 acres; $5 million of improvements were made and 165 new buildings put up to house a permanent wartime staff of 3,000. To help the morale of these troops, the army built a chapel, theater, bowling alley, and base hospital.

Brig. Gen. Charles Y. Banfill commanded both the post and training center. The MITC incorporated a headquarters with both administrative and training divisions. The training division consisted of service, supply, and 10 specialized training sections.[2] Instructors and students came from a variety of sources and military or civil backgrounds. The British provided some instructors and texts, others came

Applegate's wartime range house as it appears today at Fort Ritchie. It had been used as the base liquor store and as an office for base closure in recent years. (Melson collection.)

Applegate in the same building during the war with a selection of the training material available for instructors and students at the war's peak. An indoor range was in the basement.

from existing army schools, and some arrived from the OSS.

Graduates from MITC proved their worth in collecting and analyzing information gathered on and behind front lines throughout the world.[3] Graduates returned from the European and Pacific theaters to serve as instructors with firsthand combat experience. At one point, the instructors surveyed had commando, Ranger, patrolling, and intelligence combat service from North Africa, Sicily, Italy, the Aleutians, Guadalcanal, Bougainville, and New Guinea. There were also veterans of the Spanish Civil War, guerrilla operations in China, and fighting in Albania, Greece, and Crete.

Classes numbered between 300 and 500, with a ratio of one officer to six enlisted men broken up into 35-man training platoons. Courses consisted of eight weeks of intensive practical instruction divided into two phases. The first phase was five weeks of instruction in basic combat intelligence procedures. The second or advanced phase included

Applegate briefs the Military Intelligence Division's chief, Maj. Gen. Clayton Bissell, while MITC commandant Brig. Gen. Charles Y. Banfill and training director Col. Shipley Thomas stand at right.

instruction in the analysis of aerial photography, interrogation of prisoners, translation of documents, terrain information, signals intelligence, and counterintelligence.

Students specialized according to aptitude displayed during the course. High among talents desired was fluency in foreign languages. Sufficient instruction was given in public relations, censorship, communications, estimates and dissemination, and security to ensure graduates would have the basic concept of each in the conduct of military operations.[4] German and Japanese weapons were used at the camp for training purposes, including company-size units dressed and equipped as enemy soldiers. A mock German village was built and the post theater was used to stage simulated fascist political rallies.

In December 1942, Applegate was named chief of Training Division Section VIII, Combat Section—"Section Eight." This position he held throughout the war, progressing through the ranks from captain to lieutenant colonel as his responsibilities expanded. The section first consisted of eight officers and 17 enlisted men, organized as headquarters, instructor, supply, weapons, and range maintenance personnel. Combat Section tripled in size before the war's end.

Close-combat instruction was part of the various eight-week basic and four-week advanced courses offered at MITC. Time available for close combat and weapons instruction was as little as 14 hours at first. Applegate soon expanded this to a 24- to 26-hour block of instruction for close combat, which covered "man-killing" to "instill self-confidence and an offensive spirit in the student, and to develop his natural fighting ability with and without weapons, so that he will be able to kill his opponent, quickly and effectively in close-quarter combat." The goal was:

> . . . to instill the self-confidence and fighting ability so that he can protect himself and the army's investment in him. As the intelligence man, by necessity, carries few weapons and little fighting equipment, he must be trained to utilize any weapons which come to hand and be able to improvise and substitute enemy

Joint aspects of the MITC were based on interest and support from allies, in this case the British army's Major General Davidson and Colonel Cooper are rendered honors with the camp staff in 1943. Applegate is at left.

The typical audience at various intelligence courses for the Combat Section was a mix of officer and enlisted students, here considering the use of hand grenades.

At Camp Ritchie, lecture and demonstration were followed by application in the use of various weapons, including grenades.

Individual training was combined with more ambitious exercises that required the support of school troops and facilities, in this case a prisoner of war handling demonstration that included processing and security measures.

> equipment when he is projected into unexpected combat. Combat possibilities and situations which might confront him and the techniques he uses are realistic and practical due to the constant contacts with the theater of operations.[5]

Section Eight provided weapons, physical conditioning, and close combat and patrolling training for officers and men. Essential to this was a collection of Allied and enemy weapons, a five-lane combat firing range, an indoor combat firing course, an infiltration course, a stalking course, trainasium, and combat town.[6] The subject list (Document 5) compiled from courses during the war indicated the depth and breadth of material covered.[7]

The background and volume of students honed instruction to mass-production proportions as more than 14,000 personnel from all services were exposed to the distinctive Applegate style of close combat. A former graduate of the OSS Area B experience was reassigned to the U.S. Army and sent to MITC for training. He recalled: "I learned the army way of doing things, quite different from the individualistic OSS approach."[8]

The official history of the MID and MITC described Section Eight in detail after the war. All training methods and procedures were developed locally "to meet specific combat demands and conditions." Because of the practical nature of the subject, numerous illustrations, charts, and posters were developed, including the training films *Close Combat* and *Foreign Small Arms*. Special ranges and instructional methods were developed to make the student proficient with pistol, revolver, or submachine gun in all kinds of light, terrain, and battle conditions "with and without the use of sights."[9]

Applegate's range house office, including training aids and instruction material as published in this book. Note slogans for both armed and unarmed close combat on the walls.

A complete machine shop was available at Camp Ritchie to maintain enemy weapons and parts. It also provided the means by which to make technical improvements or evaluations.

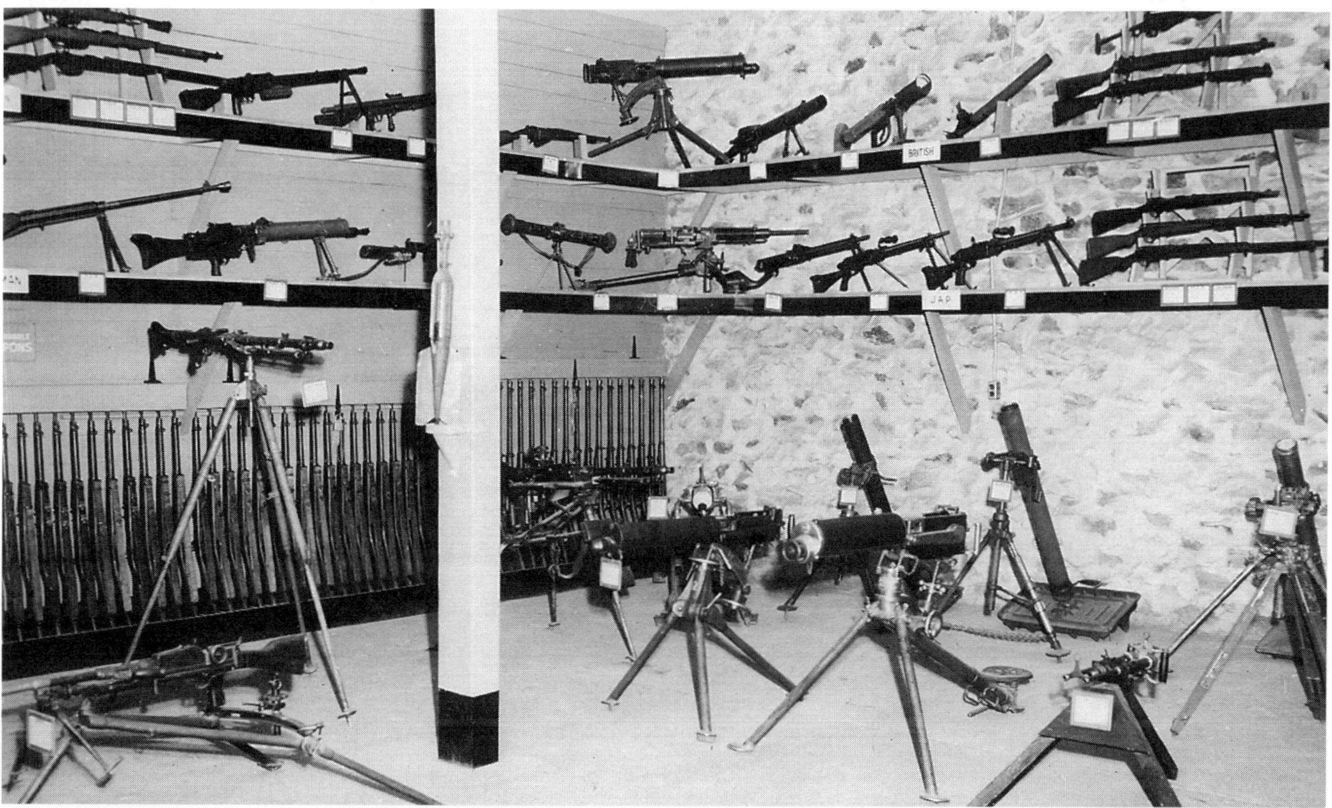

The MITC display room included examples of Allied and enemy firearms for use by instructors and students. In the foreground are German, British, American, and Italian machine guns. At left are American rifles and American and German mortars. On the lower shelf are enemy weapons, with Allied small arms on top. The collection is now at the U.S. Military Academy at West Point.

Combat Section facilities included classroom and training aids for a variety of instruction.

A sand table for use with mapping and patrol instruction. Devices like this allowed students to visualize the actions discussed.

The training aids section produced artwork and posters to reinforce lessons.

Applegate wrote that the purpose of his assignment was:

> . . . to organize and direct the Combat Section of this new center, where intensive close-quarter armed and unarmed combat instruction was the training mission. Over the ensuing two years, over 10,000 trainees from all walks of life and with many foreign language capabilities were processed there. During this time my instructional staff and I were exposed to trainees who, because of their various ethnic backgrounds, had a great deal of experience in knife and gun fighting. They were from all ethnic backgrounds, of all statures, with all sizes of hands, and with various motivations.[10]

Applegate went on to describe the sequence of training as it affected his course of instruction.

> I still have the course-of-instruction [presented in this book]. Basic MITC training followed that described. The various specialists had instruction tailored to their needs. At Ritchie we refined the moves to be taught quickly to a large number of trainees without prior training and got it down to a science. At the MITC we had resources, continuity, and documentation. The rotation of instructors and students to actual combat ensured feedback as to what worked, or did not. We could screen the students for a variety of expertise in close combat. I had no preconceived solution and it was a continuing process of education—nothing was engraved in stone.[11]

And as noted earlier:

> . . . we kept refining on what the British had been doing but still using the same principles. I sent not only trainees into combat, but instructors, who would go out on special assignments and then come back and tell us if our techniques in shooting, knives, strangulations, or bare hands, or whatever, worked. If it didn't work, we threw them out.[12]

Applegate and Section Eight worked with numerous other army commands, armed services, and government agencies with firearms and close combat, including the OSS; FBI; the Military Police School, Fort Custer; Infantry School, Fort Benning, Georgia; and the Artillery School, Fort Sill, Oklahoma. This was for liaison or cross-training and validated the Section Eight approach. Applegate recalled: "I met both Biddle and Styers at Quantico and witnessed demonstrations with bayonets. Biddle was a dilettante and showman. Both used a duelist approach that was bullshit and not based on practical experience like that of the British."[13]

Applegate made the transition from art to science between 1943 and 1945 with Section Eight solutions to close combat, the results of which brought Fairbairn and Sykes to a greater audience and amplified their efforts more than any other source. Applegate's own articles and book conformed to the same topic headings as the OSS course outlines. Emphasized was surprise through stealth and speed. Applegate brought their components together in a variation that stressed always keeping something between you and your opponent: shoulder and handguns, edged weapons, blunt instruments, improvised weapons, kicks, blows, throws, and holds. He felt that the last thing you wanted to do was grapple with an opponent.

By now, Applegate had conclusions about close combat, based on extensive research and experience that continued to evolve as long as Section Eight was in business. Some illustrative comments by Applegate about the interrelated subjects of gun, knife, and hand-to-hand fighting follow.

On gunfighting:

> Various experts, such as FitzGerald of Colt, Peret of Remington, and Askins of the

> Border Patrol, have advocated its adoption by all law enforcement agencies. Aside from immediate circles around these individuals and in some federal agencies such as the FBI, little thought has been given and little has been accomplished in teaching a man how to use his handgun without the use of sights in offensive combat. This method has been called by various names, such as body pointing, instinctive pointing, or finger pointing shooting. The Shanghai police, under the instruction of W.E. Fairbairn and E.A. Sykes, are the only ones who adopted it extensively and successfully in combating the criminal elements. Certain units of the British and American forces have been so instructed with good operational results.[14]

And:

> There has been little attempt on the part of the various police organizations and other groups using the handgun to develop practical ranges simulating combat conditions as close as possible. Most law enforcement agencies do have a course which involves running a certain distance, firing the revolver first with the right and then with the left hand, using double action and various firing positions. They are also given a certain amount of rapid fire double action shooting at bull's-eye targets and other similar firing tests. Such tests are good from a practical stand-point, but the biggest difficulty is that the officer or shooter usually fires the course only once or twice a year and is then permitted to return to target shooting. The practical shooting is not emphasized and practice in its use is not done either because of lack of interest on the part of the individuals running the shooting program, or because of the familiar old bugaboo of limited ammunition allowance. Too many organizations are more interested in competitive target shooting between groups and in collecting trophies. A combination of both types of shooting should be adopted. Good combat firing pays off not in trophies but lives. A few hundred rounds in practice in practical shooting expended over a regular period under conditions as close to those of combat as possible would save many lives of enforcement officers and would cause many more successful conclusions in gun battles between the criminal element and law enforcement agents. This same thing applies in military channels. Our soldiers spend too much time on range shooting at targets and not enough time shooting at man-sized silhouettes on practical combat courses.
>
> Camp Perry had a practical range of the type mentioned in years past under the name of "Hogan's Alley," and also one known as "Swedes in the Weeds." These ranges, however, were sidelines of the main event at Perry which stressed marksmanship on bull's-eye targets.[15]

Advanced students covered the use of the pistol and submachine gun at close quarters without the aid of sights in all light and combat conditions, with live ammunition on an indoor combat range. Popularly known as the "House of Horrors," Building A5 consisted of a large basement containing lifelike targets and other realistic combat implements. The students were given 24 rounds of ammunition, a pistol, and a fighting knife, and sent through the course. A series of targets and other realistic combat situations occurred. Most of the action took place in the dimly lit or totally darkened sections of the basement, with sound and small-arms effects being used.[16]

Applegate related that the MITC House of Horrors practical handgun range developed after a two-year period, during which time "several thousand handgun shooters, of all degrees of training and experience, fired over it." A review of these statistics led to some conclusions worth presenting.

> 1. That target shooting proficiency alone is not enough to equip the average man for combat, where the handgun is the primary weapon.
>
> 2. That the instinctive-pointing technique of combat firing is the best all-around method of shooting the handgun without the aid of sights.
>
> 3. That this type range is a reliable test of the combat effectiveness of all known techniques of handgun shooting without the aid of sights.
>
> 4. That there must be greater appreciation, by most training officers, of the physical and psychological effects of combat tension upon the handgun user. In addition to the changes in established techniques which were demonstrated, those shooters who were psychologically unsuited for combat or who had the wrong kind of temperament were discovered.[17]

Applegate and Sykes corresponded about firearms, night sights, and the development of silencers to suppress the sound made by .22 and .45 caliber pistols and the 9mm Sten submachine gun. Blueprints and examples were provided to support American efforts, and Applegate felt that this was an area the British had a head start on from pre-war civilian efforts. Sykes was suspicious of the practicality of suppressed automatic fire, which, though almost noiseless, "if it is heard it is absolutely unmistakable."[18] Single shots were preferred, particularly when used in conjunction with a background of foliage or terrain or in the open. Anything in between reflected the sound back.

On knife fighting:

> We will make no attempt to go back into the history of the use of knives and the various types which have been developed and used successfully in the past centuries. At this date, little has ever been written concerning the use of a knife for close-in fighting, and in most nations in which a bladed weapon is used, little has actually been done in instructing in its use. The knife has been considered merely a weapon characteristic of that particular area [with] each individual using it as he saw fit.
>
> Professional fencing instructors have lately endeavored to lay down programs for the training of individuals in knife work, but most of them visualize a situation from the fencer's viewpoint, in which two men approach each other from a distance with drawn knives. Thus they have tried to develop a system of knife fencing instead of close-in knife fighting.[19]

Also:

> In the present conflict, the fighting knife has two main uses, one as a reserve weapon to be used when all else fails, and the other for specific missions, such as assassination, sentry killing, or in any situation where silence and quick killing efficiency is desired. That it is important as a major weapon for troops has lately become more evident . . . although little definite instruction in its use seems to have been given to the troops carrying them.[20]

In later years Applegate commented:

> One-on-one knife fighting incidents—where both men are

> forewarned, similarly armed, and engaged—are extremely rare. Generally, knife confrontations should be over in seconds when the trained knife fighter is pitted against an untrained, or unprepared, opponent—especially when the element of surprise is present.[21]

The knife was emphasized by Applegate "as a reserve weapon to be used when all else fails" or for specific tasks "where silence and quick killing efficiency" was needed, such as sentry killing. Knife throwing was discarded because, on considering "the agility of a military target, heavy clothing, and the fact that if you miss you are without a weapon," it was evident that it was impractical.[22]

Some time was spent at MITC addressing the relative merits of edged weapons. The preferred Fairbairn and Sykes design had a handle or hilt length of ideal proportions, roughly five inches from the end of the butt to the cross guard. The blade tip was approximately five and a quarter inches from the cross guard. The point of balance was roughly one inch from the guard toward the butt. The handle was knurled to give a good grip, and the small knob on the end made it easy to pull from the sheath.[23] This design of the favored fighting knife contributed to its maneuverability when properly held and, in general, "knives with spikes on the butt, brass knuckles for the hilt, and any other additions are not too practical."[24]

As the war continued, modifications were proposed as part of the ongoing production of both fighting and utility knives. Applegate reported to the U.S. Army Infantry Board that both fighting and utility knives were justified because the existing general purpose weapons were "awkward and ill-adapted to strokes of the slashing or chopping type."[25] For field use, changes to the original fighting knife were desired: it was prone to

Motivation at the wartime MITC was straightforward: kill or get killed.

Street fighting and raids were practiced in a "combat town" that represented a German village.

The indoor five-lane firing range available for individual instruction in point shooting.

The range in use by individuals working on handgun and submachine gun point-shooting methods.

break at the tip or base of the tang from torsion or by hitting equipment and even uniform buttons. The metal handle caused gripping problems under cold or wet conditions.

Applegate and Fairbairn persisted in their mutual interests in knives and knife fighting. Even after the adoption of the U.S. Army M3 trench knife, Applegate continued to use and teach with the Fairbairn-Sykes fighting knife as produced in America for the OSS, or as available in Great Britain. The smatchet was modified into the Fairsword in 1944, and a new fighting knife design was seen by 1945. The Fairsword was a utility knife for general purpose and close combat use "superior" to the various smatchet models, and would "permit every possible type of blow being made." The Fairbairn-Millerson fighting knife enhanced grip, balance, and blade durability. While both were better, they did not justify the cost of production over existing stocks.[26] Applegate offered his expertise to the famed Orlando, Florida, knife maker W.D. (Bo) Randall after seeing an early copy of Randall's booklet on knife fighting. In February 1944, he sent duplicates of training material and the prototype of an improved fighting knife. Applegate worked with Randall to make this knife for private sale, but the $19 price was too high for most soldiers at Camp Ritchie.[27]

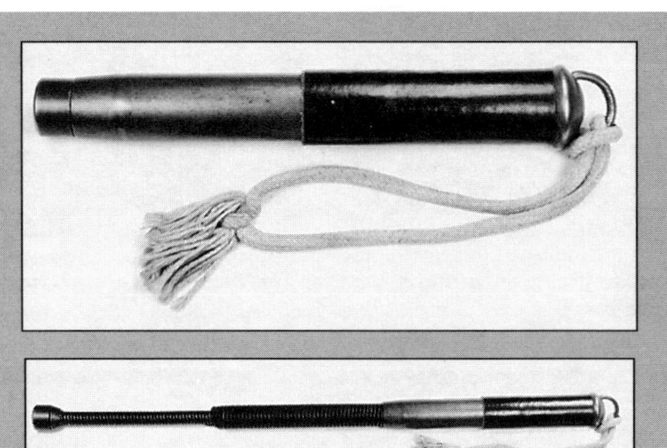

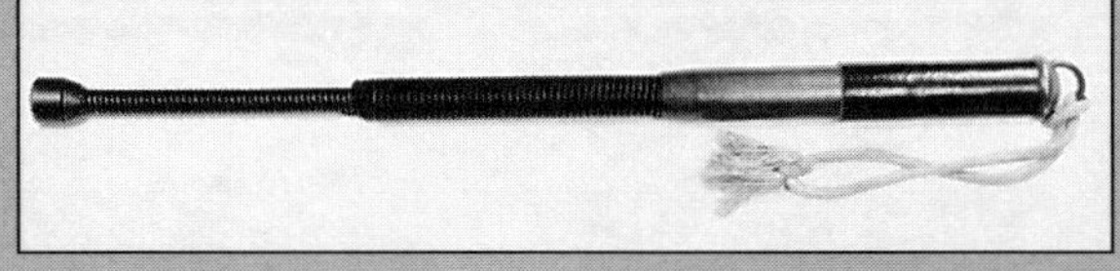

The spring cosh produced for the OSS was based on British designs. A close-quarter weapon favored by Fairbairn, Applegate felt it was too deadly for police work but not practical for combat. (Photos courtesy of John W. Brunner.)

The U.S. Army M3 trench knife and later bayonets could all be used with this knife fighting system under most circumstances. Currently effective designs include the original Fairbairn-Sykes knife, as well as the modern Randall Model 2, Gerber Mark I and II, Kershaw Trooper, Al Mar Shadow, some Ek models, and the various Applegate-Fairbairn combat knife configurations.

On miscellaneous weapons: the spring cosh was criticized by both the British and Americans as being subject to the same countermeasures as any form of raised arm attack. Applegate felt it was "the most vicious type of blackjack known" developed for underground warfare, but not suitable for police use "since it is intended to maim or kill, not stun."[28]

On disarming: techniques that allowed an unarmed man to overcome an armed assailant were given a lot of work and practice by Section Eight at MITC with pacesetting results. Previous efforts by Fairbairn, Biddle, and the FBI were considered and analyzed. The conclusion Applegate reached was that, in most cases, individuals were only trained in methods to "get the gun away from the opponent, leaving them still in a position of having to subdue him when and if this takes place. I have always felt that disarming should be incorporated simultaneously with a devastating attack on the man with the gun."[29] In the case of OSS and SOE operatives this was a vital subject, as in most cases they were taken into custody at road blocks or security checkpoints and they were unarmed. For military intelligence personnel, this was important as well in order to safely subdue and handle prisoners of all sorts.

On hand-to-hand fighting:

> With the advent of World War II, public interest, as always in time of war, had been directed toward fighting and methods of combat. The desire of the American soldier and the American common man for knowledge of fighting techniques has greatly increased.
>
> Throughout the country today, numerous articles are being written

for publication in our magazines and newspapers, and illustrations are being printed in the rotogravure sections of leading dailies of so-called "rough stuff" and underhand tactics. Throughout our armed forces various schools of instruction and courses are being given by individuals who are qualified along orthodox lines and in many cases have had a great deal of police or other restraining types of man handling. The biggest trouble is that no one has outlined a uniform system of instruction or a constructive training course for unarmed fighting.

All such combat should be instructed on the theory of whether or not it would be useful to a man after he had lost his weapons. The knife is the ideal weapon for close work and its use should be encouraged and adopted by all units of our armed forces because of its deadly and silent effectiveness.[30]

Hence the reason for a simple type of instruction with a great deal of emphasis on the few elementary methods which can be easily and instinctively used in combat. We shall endeavor to reduce it to its lowest common denominator. We do not want to make a professional out of the average individual, but rather to teach a few simple tricks which he can learn in a few minutes and use after practice.[31]

This type of instruction teaches a man to fight and kill without the use of firearms, knives, and other lethal weapons. It is designed for use when those weapons have been lost, which should be avoided at all costs, or when the use of firearms is undesirable for fear of raising an alarm.[32]

The principles of unarmed combat are largely those of judo, various other styles of wrestling, boxing, Chinese boxing, self-defense devices, and rough and tumble tactics. The importance of this type of combat lies not alone in the extreme offensive skill which its students can achieve, but also in the fact that any man, regardless of size or physique, once well trained in this technique, has a supreme self-confidence in himself and his fighting abilities, which he could not achieve in any other way.[33]

In this type of combat, we hit, chop, thrust, poke, or kick vital points on the opponent with the fist, the elbow, the knee, the toe of the foot, and the heel and edge or palm of the hand.[34]

Brigadier General Banfill and Col. Shipley Thomas noted Applegate's wartime contributions to the field through his professional contacts and writing. Beginning in 1943, Applegate's experience with close-combat material appeared in *Infantry Journal* and *American Rifleman*, and he published his classic manual of hand-to-hand fighting, *Kill or Get Killed*. Sykes felt that: "It is the first book, as far as my knowledge goes, to deal with the subject on sensible lines and without doubt will replace what I can only describe and always think of as peace-time methods."[35] The intent was to make the information available to a wider audience in the armed forces, as well as for Applegate's concern for American law enforcement, because "after the present conflict, as after no other war, the world will be faced by criminals who will take advantage of their military training" to carry out outrages in the post-war world.[36] Along with extensive lesson plans and technical reports, the text *Specialized Training in Foot Reconnaissance* was drafted and published after the war. Both *Kill or Get Killed* and *Scouting and Patrolling* are still in print more than 50 years later.

By 1945, Section Eight expanded to its most advanced form, what Applegate chronicler William L. Cassidy described as "the most elaborate training and research center ever to appear in the annals of close combat."[37] In addition to instruction in armed and unarmed close combat, instruction enlarged to included prisoner handling, observation and patrolling, and house raids.[38] Some 27 hours of material were covered in the basic course for all students, with 43 additional hours presented in an advanced course to counterintelligence personnel, and a 79-hour scouting and patrolling package for terrain intelligence specialists. When it was all over, Applegate recommended 64 hours to cover this same program; 40 hours devoted to close combat and 24 hours to weapons training.

Brigadier General Banfill considered Applegate "to be one of the best qualified men in the armed forces to train and advise in those subjects in which he has specialized here at Camp Ritchie": close combat; combat use of American, Allied, and enemy small arms; raids and street fighting; and scouting, patrolling, and observation.[39] Colonel Thomas, the MITC director of training and Applegate's immediate boss, wrote that this was due to "his own abilities and unique opportunity while stationed here, to train, study and devise methods and techniques that are the result of recent actual combat experience."[40]

Perhaps the greatest compliment of all came from Eric A. Sykes, who wrote that Applegate's approach was "clear, practical and, above all, what is wanted."[41] After the war, William H. Jordan of the U.S. Border Patrol (a U.S. Marine and noted shooter) asserted that the "techniques of hand-to-hand fighting worked out by then Capt. Rex Applegate and his staff were taught to thousands of men going into combat. Many of these returned to verify the rightness of these techniques or to give information by which they were corrected or refined. They had to be taught to be tougher, meaner, more efficient and more merciless than the enemy if this country was to survive."[42]

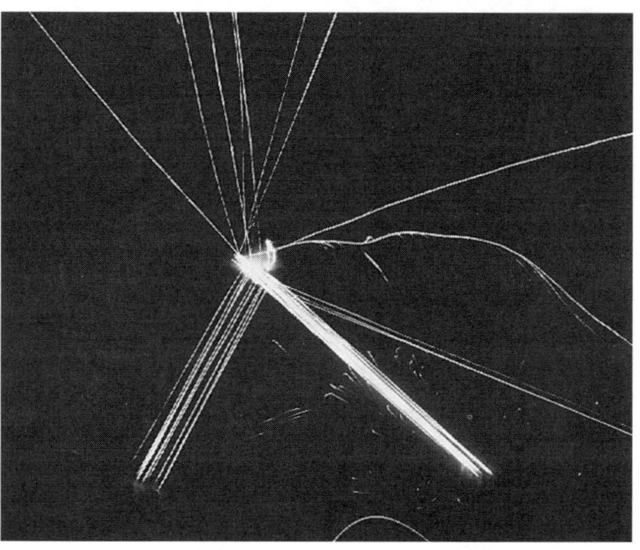

Tracers indicate the cone of fire for the machine guns, which provided visual evidence of live fire and the need for proper action in combat. Demolition charges simulated artillery and mortar fire.

Daytime weapons firing demonstration to show the sound and effect of Allied and enemy small arms. The range house provides the background to the audience.

A night exercise conducted for battle indoctrination that included actions illuminated by aerial flares.

Students realized that movement under fire was best accomplished low to the ground, regardless of what they might have to crawl through. These methods, pioneered at MITC, are now standard for most general and special purpose forces.

Movement under machine gun fire included American and German weapons firing live ammunition over the heads of low-crawling students.

COURSE OF INSTRUCTION, MITC BASIC AND ADVANCED
[Document 5; May 1945]

Period	Hours
I. Introductory Lecture	1
II. Balance Demonstration and Wrist Release, Kicks, and Kicks as Coup de Grace	1
III. Edge-of-Hand Blow and Chin Jab	1
IV. Come-Along (Wrist and Arm Lock) and Flying Mare	1
V. Wrist Throw and Pushing Counter	1
VI. Front Strangle and Jap Strangle	1
VII. Sentry Killing (Stick, Garrote, Knife)	1
VIII. Observation	1
IX. Range Estimation	1
X. Patrol Procedures	1
XI. Formations and Ambushes	1
XII, XIII. Combat Patrol Formations	2
XIV. Pressure Points, Stick Technique	1
XV. Ties (Hog and Shoe Lace)	1
XVI, XVII. Disarming	2
XVIII. Knife and Knife Defense	1
XIX. Street Fighting	1
XX. Foreign Weapons Lecture and Firing	1
XXI. Foreign Rifles	1
XXII. Care, Functioning and Firing of M1 Carbine	1
XXIII, XXIV, XXV. M1 Carbine Firing	3
XXVI, XXVII. Night Battle Indoctrination Course	2
XXVIII, XXIX. Observation Demonstration	2
XXX, XXXI. Observation Practical Examination	2
XXXII. Care, Functioning and Firing of .45 Cal Pistol	1*
XXXIII-XXXVIII. Combat Firing .45 Pistol	6*
XXXVIV. Care, Functioning and Firing of .38 Cal Pistol	1*
XXXX-XXXXIII. Combat Firing .38 Pistol	4*
XXXIV. Firing Enemy Small Arms	1*
XXXXV. Care, Functioning and Firing of M3 Submachine Gun	1*
XXXXVI. Firing M3 Submachine Gun	1*
XXXXVII, XXXXVIII. Tactical Raid Demonstration	2*
IL-LIII. Patrol Procedure and Night Patrolling	5*

*CIC Material

ENDNOTES

1. Melchior, *Case by Case*, 27.
2. Camp Ritchie, RT701. Also William L. Cassidy, *The Complete Book of Knife Fighting* (Boulder, CO: Paladin Press, 1975), 20-23; *Basic Manual of Knife Fighting* (Boulder, CO: Paladin Press, 1978); and *Quick or Dead*, 101-110. Cassidy pioneered the story of Fairbairn, Sykes, and Applegate, but is open to challenge for interpretation and facts.
3. MITCVIII, Purpose and Scope of Course of Instruction in Combat Section VIII, 4 March 1943, 1-2; Kathy Fotheringham, "History of Fort Ritchie" (Public Affairs Office, n.d.); Kathy Fotheringham, personal communication, 18 August 1995.
4. Kathy Fotheringham, "Camp Ritchie During World War II (Public Affairs Office, n.d.), 7-8.
5. MITCVIII, Purpose and Scope, 1.
6. "A History of the Military Intelligence Division," 7 December 1941-2 September 1945 (Washington, DC: MID, 1946), sections titled "Instructors," "Trainees," "Curricula," "Training Methods and Procedures, " "Training Doctrine and its Sources," and "Close Combat Section"; Camp Ritchie, RT704, 705. Applegate was asked to indicate the location of facilities on a current Fort Ritchie map: "The area by the flagpole and parade ground was the rifle range. The battle course and basement of the range house were there. The German village was up the valley. The indoor range was out in back. A lot of the World War II buildings were torn down."
7. Camp Ritchie, RIT704, 705; Course of Instruction in Combat Section VIII from 8 February 1943, 4 March 1943, 9 September 1943, and May 1945.
8. Melchior, *Case by Case*, 25.
9. "A History of the Military Intelligence Division," "Close Combat Section."
10. Rex Applegate, *Combat Use of the Double-Edged Fighting Knife* (Boulder, CO: Paladin Press, 1993), 1-2; Applegate, TREXPO, 7.
11. Rex Applegate, personal communication, 30 October 1995.
12. Applegate, TREXPO, 8.
13. Rex Applegate, personal communication, 30 October 1995.
14. MITCVIII, Hand Gun, 3.
15. MITCVIII, Hand Gun, 13.
16. Kathy Fotheringham, personal communication, 18 August 1995; RT704, 705; "A History of the Military Intelligence Division," "Close Combat Section."
17. Applegate, *Kill or Get Killed*, 279.
18. Eric A. Sykes, personal communication, 1 March 1943.
19. MITCVIII, "The Fighting Knife," 22 January 1943, 4.
20. MITCVIII, "The Fighting Knife," 4.
21. Applegate, *Combat Use of the Double-Edged Fighting Knife*, 3.
22. Rex Applegate, "Knife Fighting," *Infantry Journal*, December 1943, 46-47.
23. Applegate, "Knife Fighting," 49.
24. Applegate, "Knife Fighting," 49.
25. Rex Applegate, "General Utility Fighting Knife," n.d., 1.
26. OSS, "Final Report, Knives, General," 24 September 1945 (NARA); Yeaton, et al., *Commando Knives*, 106.
27. Charles Karwan, personal communication, 13 September 1997; Robert L. Gaddis, *Randall Made Knives* (Boulder, CO: Paladin Press, 1993), 73-75.
28. Applegate, *Kill or Get Killed*, 318-319; "Silent Killing," 30 June 1942, 7.
29. Rex Applegate, personal communication, 1 December 1997.
30. MITCVIII, "Unarmed Offense," n.d., 2-3.
31. MITCVIII, "Unarmed Offense," 4.
32. MITCVIII, "Unarmed Offense," 5.
33. MITCVIII, "Unarmed Offense," 5-6.
34. MITCVIII, "Unarmed Offense," 7.
35. Eric A. Sykes, personal communication, 22 September 1943.
36. Rex Applegate, *Kill or Get Killed* (Harrisburg, PA: Military Service Publishing, 1943), iv.
37. Cassidy, *Quick or Dead*, 101-102.
38. MITCVIII, "Basic and CIC Course of Instruction," May 1945, passim.
39. Charles Y. Banfill, personal communication, 30 December 1944.
40. Shipley Thomas, personal communication, 28 December 1944.
41. Eric A. Sykes, personal communication, 22 September 1943.
42. Bill Jordan, Introduction, *Kill or Get Killed*, by Rex Applegate (Boulder, CO: Paladin Press, 1976), vi.

CHAPTER FIVE
MITC: QUICK OR DEAD

Documents in the next three sections comprise the book's core, for they show a system of close combat at a distinct stage of development from the previous efforts by other organizations (SMP, STC, SOE, SIS, and OSS). Applegate took what was learned from them and, with the MITC Section Eight staff, refined it into a useful program for American students.

MITC handgun display to indicate the variety of Allied and enemy weapons that might be encountered or used overseas.

Far Right: These were some of the automatics, the Colt .45 (top) and the Browning 9mm Hi-Power (bottom).

Right: Applegate shows the "pump handle" ready position used in point shooting the pistol. This includes both eyes on the target and the instinctive crouch prior to bringing the weapon up to eye level and firing using a convulsive grip.

Below: Specific small arms used for MITC close-combat training included Smith & Wesson .38 revolvers, Colt .45 M1911A1 automatic pistols, U.S. fighting knives, Thompson .45 M1928 submachine guns, and .30 M1 carbines.

The work documented at MITC falls into roughly three distinct revisions: 1) that from January 1943 was subsequently published in magazine articles and the July 1943 edition of *Kill or Get Killed*; 2) that from the 9 September 1943 course of instruction is published here and presents the clearest exposition of what was taught during the war; and 3) that from early 1945 was used for the fourth and later editions of *Kill or Get Killed.*

In addition to close-combat training, Applegate worked on a variety of specific issues of interest to MID and others. A selection of these files are included because of historical and continued relevance.

HANDGUN OFFENSE
[Document 6a; 14 March 1943]

1. Introduction

"Fear no man, no matter his size. When trouble threatens, call on me—I'll equalize." Judge Colt

It has been said that all men were not born equal—Samuel Colt made them that way. The fast handling of the handgun in pioneer days put the little man on even terms with the big fellow. In the pioneer West, where every man was his own law, the Colt, which was called in popular slang "the great equalizer," was indispensable. The highest premium of all—life—was often the reward for skilled gun fighting.

The gun fighters of our early days naturally were not equipped with weapons of the mechanical and scientific perfection of the present day, but the principles which were used and the element of skill in shooting were identical to those of today.

Today no other nation in the world has adapted the handgun to general use in sports and police work as has America.

Use of the Handgun: The purposes of the handgun in the present age are two: for use in sports and for use as a weapon of combat. The type of shooting and instruction involved in these two phases should be as different as black is from white. While the high degree of skill attained by expert target shooters is to be admired, it is of little use to men in close combat.

We are concerned only with the use of the pistol as a means of offensive combat, and all comments hereinafter will be based solely on making the student familiar with the handgun, primarily as a means of offense, secondarily as one of defense.

To be able to hit a black dot at a given number of yards is not nearly as important as [the ability] to hit an enemy before he gets you. The desired goal is to ingrain in the shooter a supreme sense of confidence in his ability to use his gun so as to get there "fustest with the mostest" lead.

It is a matter of record that the majority of shooting frays between individuals take place at a distance of not more than 20 feet. Consequently, the man who can use his weapon quickly and accurately from any position without using the sights is the one who will stand the best chance of not going out feet first.

In handgun offense the circumstances are entirely different from those of target shooting. Here, speed in firing, confidence of the shooter in his weapon and in his ability, and practice under conditions which approach those of actual combat firing are the important factors.

2. Shooting by Instinctive Pointing

This method of firing is done with the body in a crouching position. The arm is fully extended, although the elbow may be slightly flexed. The grip of the hand and the weapon is tight—almost convulsive.

The crouch is used because it has been found that when subject to enemy fire, men will instinctively assume this position. With the arm extended, control of elevation and windage is more easily effected and the necessity for the extreme amount of practice in learning to shoot accurately with the gun resting on the hip is eliminated. The grip on the pistol or revolver is extremely tight.

This is also a combat condition, because in the midst of battle excitement, a man instinctively grips his weapon in this manner and certainly does not take time to hold his breath, line up the sights, and squeeze the trigger.

The best system in firing which practicability must be considered and in which the sights are not used is as follows: Body crouched, arm extended with the gun in a perpendicular line with the belt buckle and the eyes. It is far enough forward from the shooter's crouched position so that he can see his weapon and the target at the same time and can thus easily line up the two to get accurate shooting.

To achieve the desired position of the pistol it is necessary to flex the wrist a slight degree to the right, assuming, of course, that you are a right-handed shooter.

The man who shoots in this manner fires in the direction in which his body points. In other words with the wrist and elbow locked and the arm extended and maintained in the same relative position to the center of the body, he is automatically on a line for accurate work by wheeling his body in any direction and looking at the target. As the arm is at all times in locked position at wrist and elbow, the only movement will be raising and lowering the weapon from the pivoting point of the shoulder.

Use of Mirror: Practice for the student should proceed somewhat in the following manner: He should be placed initially about 6 feet in front of a full-length mirror and told to assume the crouching position. Either left or right foot may be forward.

After going into a crouch, he should raise and lower his hand, all the while looking at his image in the mirror, never once looking at the pointing finger, but only at the spot he wishes to hit. In the crouch, his body should be leaning forward, his shoulders as nearly parallel as possible.

After a short period of practice in this position in front of a mirror, he should continue the same type of practice with a gun, snapping the trigger as he raises his gun in line with the point on his reflection he wishes to hit.

Where to Aim: The best part of a man's anatomy to shoot at is his middle, because a man-stopping hit can be made in that area easily. Any hit a foot high or low or a few inches to the right or left of the body center is almost always a good one.

He should then be placed facing the mirror at right angles and given instructions to wheel and snap the weapon at the mirror image.

Due to different positions in which the feet will be at time of firing, the shooter should let his body position change by moving his feet in any natural way. Instruction in a set method of wheeling the body and moving the feet is not advisable because of the uncertain elements of terrain, ground, and foot position in combat.

Next the student should place his back toward the mirror and whirl around toward the target. At this stage the necessity for always firing with the arm extended and the gun in line with the belt buckle and the eyes, letting the body do the actual aiming at the target, is apparent.

You can then demonstrate how much better this method is by allowing him to face right or left angles from the target and in place of turning his body merely have him swing his arm from right or left toward the target.

It is easy to see that it is very difficult without turning the body to swing your arm to a new direction and maintain the proper windage for accurate firing. Usually 2/3 of the shots will be made either before your weapon points toward the target or after it has passed over and is on the other side.

The method used by some instructors involves jumping instead of a foot movement to change body direction when firing. This method is not desirable due to uneven terrain, the chance of losing balance, and direction control.

Silhouette Targets: Going back to the student who has taken his preliminary instruction in proper grip, stance, and body position in front of a mirror, after he has practiced a sufficient length of time to master the fundamentals, he may be allowed to fire at silhouettes. This expedient has been used with a great deal of success to achieve fast, accurate results with a minimum expenditure of valuable ammunition, which is often a major factor in training.

Use of Toy Gun: After the pointing stage has been mastered, the student should procure a toy gun which fires a little wooden dart with a rubber suction cup on the end. It can be used with the darts just as it comes from the toy counter, or the mechanism can be placed in a wooden dummy the shape, size, and weight of the gun the student will later fire. This toy makes it possible for the shooter to see in the mirror the exact point of impact of the bullet and also to see in the mirror his own errors.

Small BB type pistols have also been used for this type of practice preliminary to firing live rounds.

Live Ammunition: After these preliminary stages have been passed, the student should be given live ammunition firing with two-shot bursts at a distance of not more than 8 feet at a man-sized silhouette mounted against a background which will show misses. He will thus be able to see his hits.

Four of the most common errors are:

a. A loose grip on the weapon which causes dispersion of shots on the target. This can be easily corrected.

b. Failure to raise the weapon to a point where the barrel is on a parallel to the ground surface, a fault also easily eliminated by practice.

c. Failure to adopt a locked wrist and elbow in the shooting arm.

d. Instead of using the pivot of the shoulder joint alone when raising the weapon, the student may shove the arm and gun forward when firing. This causes the barrel to point downward.

After the student has mastered the feel of his weapon when firing live ammunition and can place his shots in a group no larger than the spread of an average hand, let the distance be increased to a maximum of 20 feet. At this distance a group which can be covered by the spread of two hands is good.

From this point, which has been strictly frontal firing, return to the 8-foot range and let the student practice firing at a silhouette from right and left angles, taking care to see that he makes complete body turns, changing his feet position naturally and instinctively.

A few hours' practice as outlined in the foregoing paragraphs will develop in the average individual a remarkable ability to shoot quickly and accurately.

3. Automatic Pistols and Revolvers

Any of the basic fundamentals mentioned are applicable to both revolver and automatic type weapons with a few small variations in procedure due to differences in the structure of the two types of handguns.

As the pistol or the revolver is grasped in a vise-like manner by the shooter at the time of raising his arm and firing, the structure of the weapon will obviously affect the firing because such a tight grip invariably causes the gun to be grasped along the line of least resistance and conforms to its general design.

The .45 Caliber Automatic: Outside of the North American continent the revolver is not much in use, so first the automatic or pistol type of handgun should be considered. Generally speaking, automatics fall into two classes: the U.S. Army .45 caliber, which is in a category of its own, and all other well-known types of automatics, both American and foreign.

Considering first our .45 Colt automatic, let it be said it is the finest military pistol in the world. It is much freer from jams and malfunctions, it is easily stripped, parts are interchangeable, the tolerance between moving parts is large and consequently small amounts of dirt or sand with which it might come in contact in field and combat conditions do not necessarily impair the function of the weapon.

Because of the butt construction of the .45, a tight grip such as advocated will cause the weapon to automatically point the barrel downward when the arm is raised to fire. This necessitates a slight upward cocking when firing the piece.

The tendency of the target shooter when he is forced to fire without using the sights is to shove the gun forward to shoot low.

If a shooter is instructed in this type of firing with the .45 automatic, which is frequently the case because of ammunition supply, and is then projected into a set of circumstances where he is given any other type of handgun for his own personal use, he will find that the cocking of his wrist which he had necessarily developed by practice with the .45 will not give him accurate shooting, but will cause him to fire high. As most shooters prefer to shoot with the arm extended, slightly flexed at the elbow, this wrist cocking is necessary in firing our military sidearm.

For practice it is best either to equip the .45 with an adapter, making its gripping and pointing qualities similar to those of other automatics and revolvers, or to analyze your future prospects and if there is possibility of using another type of weapon, shoot in the following manner: hold the gun in a tight grip, without cocking the wrist, arm straight and rigid, and raise the gun to a point almost on a level with the eyes.

Although the distance to which you raise the weapon is practically doubled because the elbow is not flexed and the time of firing is a fraction of a second more, a shooter trained in this manner can fire a .45 or any other type of handgun with accuracy in this pointing position. Having once learned it, the shooter will not be forced to change his wrist action in changing weapons.

Other Calibers: The Colt .32, all European 7.65mm weapons of comparable type, and the Luger (the finest [for] pointing of all automatic handguns) are all so constructed that a natural tight grip without the wrist cock and a slightly flexed elbow will cause the gun to point accurately when the arm is half raised so that the weapon is at a point midway between the belt and eye level for the firing position.

All that need be said concerning the revolver is that it points and feels natural in the same manner as the above mentioned American and European automatic weapons. Stress has been placed on the .45 because individuals instructed only in the use of the .45 handgun and forced to use some other weapons in operations will have frequent misses when firing.

Revolver versus Automatics: Much has been written on the merits of the revolver versus the automatic as a weapon of personal defense or offense. Throughout American law enforcement circles, over 90 percent of plain-clothes men and uniformed police carry and use the revolver type. The reasons for their choice are many and varied, but much can be attributed to the fact that the revolver is the historic type of handgun throughout the winning of the West.

Initially the revolver has better frame construction for swift and comfortable grip and draw. It is faster on the first shot (double action) than the automatic. It has better all-around balance and pointing qualities than many automatics and by using various grip adapters can be fitted to any type of hand.

Another reason for the use of the revolver in law enforcement has been the fact that larger calibers could be used in it than in automatics without entailing excess bulk or size. On the other hand, the automatic, due to its structural characteristics, is an easily concealed weapon. The .32 and the .38 caliber revolvers are widely used by civilians and police alike.

The widespread belief that the automatic is not a reliable weapon and is subject to jams and malfunctions is erroneous. Well-made automatic weapons given proper care will function dependably and efficiently.

Some of the real advantages of the automatic type weapon are that it is easier and quicker to reload, and after the first shot it can be fired with much greater accuracy and rapidity. In the instinctive pointing type of shooting, groups or bursts may be initially more accurate for the beginner because of the lighter and shorter trigger pull than on the double-action revolver.

Frequent Inspection: In comparing the number of malfunctions of the pistol and revolver from operational experience, in both cases when the weapon was treated with ordinary care it has given long and satisfactory service.

However, in military service and in other situations where weapons are issued and carried of necessity, and the carrier has no particular liking for the weapon and considers it in the same light as any other piece of equipment, it is necessary to have periodic inspections and checks to keep the gun at its top mechanical efficiency.

Jams in the Automatic: The majority of jams with the automatic type of weapon can be directly traced to the magazine. On close examination, you may find that the lips which hold the shell in place under the spring tension have been dented, bent outward or forced from their original position by dropping the magazine or by improper loading. Magazine springs should be treated properly and it is inadvisable to leave a magazine fully loaded over a period of years causing the spring to be depressed to its maximum.

Whenever possible, have more than one magazine for your weapon and change them frequently. Carry the spare magazine with one or two fewer shells than its capacity.

Magazines should be kept dry and should not be carried loose in the pocket where they will be subject to body perspiration, lint, dust and denting from other objects in the pocket.

Stopping Power of Various Calibers: Generally it has been considered that a big, slow-moving bullet such as the .45, which possesses a large diameter and great weight, is superior to a lightweight, faster moving bullet. The explanation for this has been that if you hit a man on the chin with your fist, all the force of the blow is transmitted to the recipient.

The big, slow-moving bullet functions in this manner. All the energy of the bullet is exhausted at the time of impact and the bullet does not penetrate and go on through. When a speedy, lightweight bullet hits a target it generally penetrates the target and sings off into space, wasting a lot of muzzle velocity and shocking power.

Although the above is generally recognized as the standard argument for the large caliber handgun, many instances are on record of such large calibers failing to stop individuals in combat, and on the other hand there are instances showing where small caliber bullets have done the job as well as any other size.

No one caliber is best for all cases. Larger calibers generally are better for man killing, but are not infallible. When speaking of the stopping powers of bullets, the human factor must be considered. The position of a man's body at the time of impact, whether he is off or on balance, the spot hit, the size of the man, his resistance to sudden shocks, his animal courage and fighting spirit all affect the stopping power of the bullet regardless of the caliber of weapon or size of the bullet.

4. Holsters

A holster should be made of the best leather and can be designed and supplied by any reputable manufacturer. Don't jeopardize your life with a cheap holster of flimsy lightweight construction.

As to the position in which the holster should be worn, that is entirely up to the individual. If you are working with gun and holster in the open, have it in a place where you can move freely, where the butt can be easily grasped, and where it can be drawn with speed and fired without unnecessary delay. If it is a concealed holster, always bear in mind that it should be in such a place that, regardless of type or state of your clothes, you can get to it with the least possible delay and unnecessary movement.

Once having chosen your spot for carrying a weapon, concealed or otherwise, do not change. Practice drawing your gun and firing daily.

5. Training Methods

One successful stratagem as the first instruction to students who are going to be [taught] combat shooting is for the instructor to line them up and, standing out a few feet in front of them, fire a foot or two on each side and above their heads to give them the effect of muzzle blast from the front, which is entirely different from that behind the gun.

If you have a group who have never previously fired weapons and you have only a couple of training hours in shooting, the following method is successful. Show them the proper stance to fire from a standing position. Let them grasp the pistol for firing in the same manner as they would in instinctive pointing type of shooting in a tight, almost convulsive grip.

In place of using a standard target, let them practice firing at silhouettes at a distance of not more than five yards. Even the poorest shot will score a fair percentage of hits on the silhouette and the result is that he feels he can hit a man if forced to and his confidence is thus greatly increased in himself as well as in the weapon.

Don't ever try to teach a man to fire a gun when only a few rounds are available by allowing him to shoot at a standard practice target, because missing the bull's-eye makes him feel that he is not handling the weapon accurately.

Use of Blanks: Blanks can be used to great advantage in quick draw and stalking work. Take two students, placing one of them on one side of an outdoor range and one on the other. Give them pistols with blanks and let them advance toward each other taking advantage of cover. This particular type of training, whether it is done in the woods, in buildings, or elsewhere, is a close approach to actual combat conditions.

The students must be cautioned to be sure that they are firing blanks and not to fire extremely close to each other where powder blasts will cause burns.

Safety Factor: Of all the types of shooting now in existence, none must be watched with more care than instinctive pointing for the safety factor. Automatically check your weapon for live ammunition each time you pick it up. Do so until the action becomes instinctive. It may well prevent accidents and in actual operations may save your life.

Practicing the Draw: To get men to practice quick draw

methods, the following has been used with some success: Pair off students who will be working together for several hours. Let them carry their guns in their holsters and proceed about any other training in which they may be engaged. Have their weapons doubly checked for safety features. (It is best to fill the cylinders with wax or by some other means which makes it impossible to place a live round of ammunition in the weapon.)

While they are proceeding about their duties, have one of them when in contact with the other give a previously agreed upon signal (such as "reach," "draw," etc.) at an unexpected time. The student receiving the command will execute a quick draw, point his gun at the one who issues the command, and pull the trigger.

Fighting Spirit: Develop by every means possible the fighting spirit. Train the shooter always to advance toward his target when firing. Place debris in his path and make him walk through it firing as he goes. Give all types of practical firing situations which involve changing hands, running, different positions, etc.

The question of how he would react in the face of firing directed toward him and whether or not his reactions would be the same as in practice has often arisen in the student's mind. The answer is in the affirmative, because practice will make his drawing and firing instinctive and he will not realize that he is actually being fired upon.

Through repeated experiences of individuals involved in night shooting with the handgun, it has been found that the shooter instinctively fires at gun flashes from his enemy. This provides a real reason for moving, rolling or getting out of the area of your gun flash the moment you fire.

If, in darkness, a gun flash looks oval (the shape of a football), you will know that your enemy is firing directly at you from your front. If, on the other hand, the gun flash is a streak, you will know that the shooter is firing from an angle and that you are not directly facing each other.

Reloading: Teaching a man to reload his weapon quickly is often neglected. This can be attained only by practice and by establishing competition between students to see which one reloads fastest. This practice should be done slowly at first, and the tempo speeded up after proficiency has been acquired. This practice should be done in pitch darkness as well as in daylight.

Students should be instructed in two-handed firing for long, deliberate sighted shots. They should be shown the proper method of prone firing in a two-handed rest position.

The student should also be instructed and allowed to practice firing with his left hand (or the hand not normally used). Often the right hand is put out of action and it is possible for the man to use the gun at close quarters with the other hand.

Practice should also be given the student in firing under all types of light conditions, including complete darkness where he will fire at sound. He should also be taught to fire from a prone position on the floor, arm extended, and to roll after each burst of shots.

In all this actual combat firing, instruct the student to fire in bursts of two. Observation of the hits from bursts of two shots on the silhouette will show that average bursts travel horizontally. The shots will automatically be spaced from 6 to 8 inches apart. Firing in bursts of two gives additional hitting probability.

An MITC student demonstrates the correct close-combat firing position used for shoulder weapons, in this case the Thompson submachine gun.

Two MITC students practice with shoulder and hand weapons, the firing position at left and the ready position at right. Silhouette targets were used instead of more conventional bull's-eye targets.

6. Firing Submachine Guns

The Thompson, Reising, British Sten and all other Allied types of submachine guns can be fired by the instinctive pointing method with great effect up to a distance of 40 yards. The principles which were mentioned in firing the pistol in this type of shooting are generally the same when applied to using the submachine gun without the aid of sights.

If this style of shooting is once mastered, it is possible to fire any shoulder weapon, including the new [M1] carbine and standard military rifles, with surprising accuracy in close-quarter work and from the hip positions. The conditions under which the shooter fires are virtually the same. The gun will be fired from a crouch, the grip will be convulsive, and firing will be in bursts. The principle of pointing the weapon by movement of the body instead of by use of the arms still applies.

The butt of the piece should be resting tightly against the hip, held there by the right elbow and forearm, which should be pressing in toward the right side. The grip of the right hand upon the stock of the weapon should be tight. The forearm should be lying flat in the palm of the left hand, which locks the weapon in such a position that the muzzle is on the same line of sighting as the pistol in the firing position. In other words, the muzzle should be in a straight line between the eye and the belt buckle.

From this position, the gun will be automatically in line so far as windage is concerned, and with a slight amount of practice elevation will take care of itself. It is easier to fire the submachine gun or like weapon in the instinctive pointing type of shooting than it is to fire the handgun. Consequently, shooters can be trained in 50% less time and with half the ammunition expenditure.

From the crouch position, shooting in the above described manner, the eyes are roughly two feet above the plane of the barrel. This position is also easier to fire from; because the gun is locked in position on the hip and on the forearm, there are fewer points which must be watched for correction than in the handgun firing position.

Alternate Positions: There are two alternate positions which should be mentioned. The first of these is the more desirable. All principles are the same except that the shooter crouches even more over his weapon than described in the first method and his head is lowered until his line of vision is on a plane about 10 to 12 inches above the barrel. The butt of the weapon is placed under the pit of the arm instead of resting above the hip bone.

This position has the advantage of bringing the eyes closer to the line of fire of the weapon. Consequently a good many individuals can fire in this manner with more accuracy from the very first practice.

The other firing method is placing the butt of the weapon in the middle of the stomach, resting above the belt buckle, and grasping the forearm over the top instead of underneath as in the other two methods. Good results may thus be obtained but disadvantages of not having a locked position for the stock of the weapon and a somewhat awkward way of grasping the forearm are apparent.

<u>Control of Bursts:</u> Experience has shown that the weapon is best carried on full automatic control rather than on semiautomatic. In place of hugging the forefinger around the trigger and pulling back, the student should learn to trigger the gun in the following manner: let him keep the finger rigid and strike the trigger. This gives a controlled burst of from one to three shots and at the same time keeps the weapon where larger bursts may be fired without any change in adjustment.

Such things as safeties on submachine guns may be necessary in some cases, but most of the time they are placed in such a position that in adverse weather, light or other poor conditions, it is difficult to release the safety to enable firing to begin without too long a delay, which may prove fatal.

If you are entering an area where firing is imminent, have the weapon on full automatic, trigger finger inside the guard pressing firmly against the inside of the front of the guard.

Another good method of carrying the weapon prior to entering the area of immediate firing is to keep it on full automatic with the slide forward, safety off. The only movement necessary to get the gun into quick action is to strike the slide to the rear with your left hand and pull the trigger. This is much faster than trying to fuss with an awkwardly placed safety, especially under adverse conditions.

The care of the clip is important with the submachine gun as with the pistol.

Until quite recently, guns of the submachine gun type had been equipped with too elaborate sights and too much instructional emphasis has been placed on firing them from the shoulder only. Sights should be very simple, and the firing should take place in most cases from the hip position because primarily the submachine gun is a quick-fire, short-range weapon for close-quarter work.

This is especially true in jungle warfare, street-fighting tactics, and close-in work such as night raiding and commando tactics.

7. Conclusion

In conclusion, there is one thing to impress on the student above all else. This is the ability to use his weapon instinctively, without recourse to detailed thought on the matter. Further, the user must have absolute confidence in his weapon.

Drawing, pointing, and firing must become a matter of habit. This can be accomplished by proper instruction, a determination to learn, and a few hours' practice.

In short, the user's weapon must become as much a part of him as an arm or finger—when he wants to use it, it's there.

FIRING BY INSTINCTIVE POINTING
[Document 6b; 9 September 1943]

First Period

In the introduction to weapons, the desired positions are demonstrated, the fire control mechanisms are pointed out and explained as to their operation, and the nomenclature of main parts is given. Instinctive pointing as opposed to sighting and aiming is demonstrated and explained to the student. No attempt is made to go into detailed stripping or nomenclature of the piece.

Second Period

After the safety devices, the grip, and the loading and clearing of the [.45 automatic] pistol have been mastered by the student, the dry-run phase is taken up. Full-length mirrors are used and the student approaches the mirror in the stalking crouch position. Thus he can see his form and correct many faults himself. Much time should be spent on the pivot, so that any change of direction is done by pivoting the feet and legs while maintaining the fixed position of the arm and pistol, in the center of the body. For pivot practice the instructor should stand before the group and point in quick succession in different directions, and have the men pivot in the direction designated by the instructor.

Third, Fourth, Fifth, Sixth Periods

The men fire single shots at silhouette targets at about 3 yards at first and, as their firing improves, they move back to 10 yards, and then finally to 15 yards. Each man is coached in his firing by an assistant instructor. When sufficient proficiency is attained in single-shot firing, the students fire [at] the same ranges in bursts of two.

Permit students to mark targets to stimulate interest.

Seventh, Eighth Periods

Introduction and firing of .38 revolver. The same procedure is followed as that in the instruction of the .45 automatic.

Ninth Period

Dry work with tommygun. Stress position of weapon with regard to body and eyes. Show the two methods of firing weapon in instinctive pointing. Show "safeties" used in combat.

Tenth Period

Firing tommygun at close range. Ten rounds of semiautomatic.

Eleventh, Twelfth Periods

Firing tommygun in bursts. Stress importance of getting off bursts in two or three shots. Show method of "fingering" trigger. At all times stress the importance of gun position and body position.

Thirteenth, Fourteenth Periods

Firing [M1] carbine from shoulder. Mention that it can be fired using instinctive pointing method the same as with the tommygun.

Show "hunting" method of carrying weapon with sling over shoulder and show its advantage over the "GI" method.

PRACTICAL INDOOR COMBAT COURSE FOR THE HANDGUN
[Document 6c, n.d.; Original Write Up on House of Horrors]

The above floor plan is made this way simply because we had a basement to work with and this consisted of three separate compartments as shown on the drawing. The floor is dirt and all walls and pillars are covered by two inches of wood and six inches of dirt.

The procedure is as follows: The man who is going to run the course is brought into a small room at the head of the stairs, and he is told to strap on a knife and read the instructions, which are posted on the wall. He is told to wait in this room and await further instructions.

At this time, the record player is started, and, by means of

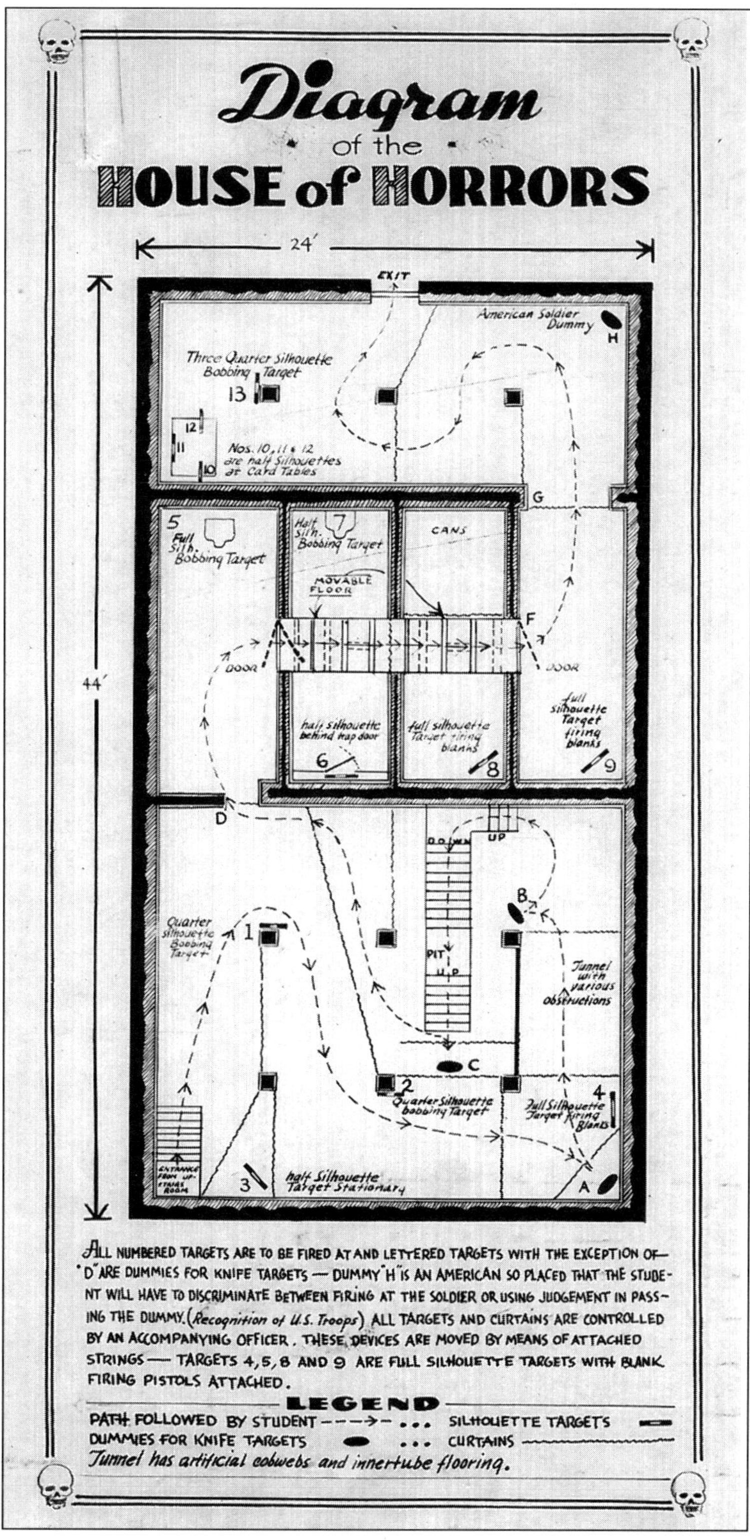

A floor plan of the House of Horrors at MITC Building A5, for which no photographs are available. This was a simulation that was designed to use point-shooting techniques under stressful conditions.

a loudspeaking system downstairs, the man is subjected to several record sequences such as German and Japanese speeches and foreign newscasts. The student, of course, cannot see the record player.

At the appropriate time, the coach calls the man from the room, guides him to the head of the stairs where he gives him any further instructions, asks him if there are any questions and hands him his pistol. At a signal, the record player commences the "This is It" sequence, and the man and coach start down the stairs. At this point are "Shots and Screams" sequences.

When the coach and student reach the foot of the stairs, target #1, which is a bobbing target concealed behind a pillar illuminated by a red light, is exposed. After firing at this, the man and coach continue around target #1 and target #2, which is also a bobbing target concealed behind a pillar and illuminated by a red light.

Target #3 is a 1/4 size stationary target and is exposed to the student's view by pulling aside a curtain. This target is also illuminated by a red light. Target #4 is next encountered and is concealed behind a curtain. This also is illuminated by a red light and when the curtain is pulled exposing it, the target, which is equipped with a blank pistol, fires at the student. At this point the student's gun, which is out of cartridges, is taken from him by the coach and the [student] is told that he will proceed through the tunnel and will be met on the other side by the coach. Before entering the tunnel, however, the coach exposes dummy "A" and the student uses the knife on it.

While the firer is proceeding from target #1 to target #4, the "Gestapo Torture Scene" sequence or the "Italian Cursing" sequence is used. As the student enters the tunnel a signal is given by the coach to the record player upstairs, and the record player ad-libs over the loudspeaker while the student is going through the tunnel.

After stabbing dummy "B," the student proceeds up the stairs onto the platform and then he descends into a pit about four feet deep. At the appropriate time, the coach, who is observing the student's progress, signals the record player, who in turn starts the "Sentry Killing" sequence.

The coach then pulls aside the curtain concealing dummy "C" and the student stabs it. Dummy "C" is illuminated by a blue light. The coach then takes the knife away from the student and leaves it on the ground, as the [student] will not need his knife after this. Immediately after the knifing of dummy "C," the record player starts the "Dog Barking" sequence.

The coach and student now proceed to the spot marked "X3" on the chart, where the coach gives the student his pistol, pulls aside the curtain and they both proceed into the second room. Here the coach may do any of two things: He may either expose target #5, which rises from the floor, or he may ignore target #5 and fire at targets #6 and #7. If he uses target #5, which is illuminated by a red light, then he has the student kick open the swinging door and fire at target #6, which is concealed in a window and is exposed by having the shutter swing back. Target #6 is illuminated by a blue light. If he ignores target #5, he has the student pull open the swinging door and exposes target #6 as described before and then exposes target #7, which rises from the floor and is illuminated by a red light.

During the firing at targets #5, #6, and #7, the "Water, Water!" sequence is used. They enter the next compartment where target #8 is located. This target is in darkness and is equipped with a pistol which fires a blank at the student as he enters the compartment. To the left of the student, tin cans are set along the wall and are rattled by the coach by pulling a string. A blue light is on the floor near the cans.

The student next enters [the] compartment where target #9 is situated. To enter the compartment he must kick open a door. This causes target #9, which is equipped with a blank pistol, to fire back at him. Target #9 is in darkness, and as the student enters this compartment the "Jap Rape" sequence is used.

Coach and student proceed to "X4" where the coach reloads the student's pistol. This position is cut off from target #9 by a curtain and is illuminated by a blue light.

While the coach is reloading the student's pistol, the "Get that American Son-of-a-Bitch" sequence is used. As the coach and student pass through the curtain into the next compartment, they are confronted by a dummy which has a knife stuck in its back, and represents a dead body. This dummy is illuminated by a green light and is not to be fired at by the student, although practically all of them do.

As the coach and student pass this dummy, the "Card Playing" sequence is started and, at the appropriate time, the curtain is pulled aside, exposing to the student's view targets #10,

#11, and #12. These targets are illuminated by two candles which are placed on the card table in whiskey bottles. When the student finishes firing at targets #10, #11, and #12, he should have two shells remaining in the pistol. As no other targets are in sight, he unconsciously relaxes and, at the psychological moment, the coach exposes target #13. This is a full-size bobbing target which has been concealed behind a pillar and is illuminated by the two candles on the card table. The student is then ushered out the rear exit and the coach returns, marking the targets and resetting them in preparation for the next man.

The targets which are fired at are made of plywood and have faces and uniforms painted on them to represent both German and Japanese soldiers. The stabbing dummies are made of salvaged fatigue clothes and are provided with a face and helmet to add to their life-like appearance. Targets #10, #11, and #12 are set around the card table and have uniforms and faces painted on them to resemble German soldiers.

This range is in the nature of an indoor combat assault course and may be constructed with scrap lumber and at very little cost. The floor plan is very elastic, and any building or basement which can be bullet-proofed may be used and the targets situated to fit any particular plan. The usefulness of the range lies in the fact that it gives men practice firing at targets under simulated battle conditions and under all types of lighting. Men will gain confidence in themselves after they have run through the course several times and discovered the high degree of accuracy which they can attain with the instinctive pointing method of shooting (which is combat firing with a handgun) in place of the standard target shooting method.

This range also brings out the fact to the student, when he initially fires, of how poorly the individual shoots at extremely close quarters when subjected to combat tension. The need for the type of shooting which will enable him to make hits without using the sights is thus clearly emphasized.

In the above described range, the student who is initially sent through before any actual practice in this type of firing will register on the average of not more than four or five hits on the targets at which he fires. Even though none of these targets are farther away than 10 feet, the majority are 5 or 6 [away]. After instruction in the instinctive pointing method of shooting in all light conditions, the student will double the number of hits at the same range after a short amount of instruction; 15 hours practice, both dry and actual firing, in this type of work will bring the average up to almost 90%.

A system of scoring in a range of this type can be made up after a few shooters have gone through and an average established.

When conducting a shooter through a range of this type, the coach should be directly behind the shooter in such a position that he can grab his shooting arm at any instant. The coach also controls the targets. As live ammunition is used and army regulations state that an officer will be present at all firing, the coach will always be a commissioned officer.

There is no end to the possibilities of this range. The limitation is only the ingenuity of the builder.

Sequences

The sequences are more fully described as follows: The sequences which are discussed are in the form of records and are played from the machine upstairs through a loudspeaking system which extends through the whole basement.

"This is It"—consists of asking the student if he has his gun and if his knife is in place and cautioning him to take it easy.

"Shots and Screams"—a burst of shots followed by a gasp and groan.

"Italian Cursing"—cursing in Italian.

"Gestapo Torture Scene"—questioning of an American-Jewish prisoner by Gestapo. Sounds of blows and German cursing.

"Ad-libbing Through Tunnel"—consists of telling man that the dummy which he encounters in the tunnel is only a dead body and that he will see plenty more like that before he is done.

"Sentry Killing"—There is a sentry! Better use your knife on him! More loud screams.

"Dog Barking"—barks, snarls, and growls by an angry dog.

"Water, Water!"—They got me in the guts. Ah! Ah! Groans. Water! Water!

"Jap Rape"—Jap talking to American girl followed by her screams and his sinister laugh.

"The American Son-of-a-Bitch"—Let's get that American son-of-a-bitch. Followed by shots, screams and groans.

"Card Playing"—several Germans bidding and arguing at a card table. Sounds of coins being dropped on table.

Safety Precautions

With which all officers who conduct students through the range are familiar.

1. All guns are loaded with .22 caliber ammunition only.

2. Instructor is to load all guns with chamber empty when given to student. Student, in turn, pulls slide to load.

3. Instructor follows man with right or left hand always on man to keep contact and prevent him from turning around. Also, [instructor is] in a position to grab man's gun in case the latter turns around.

4. All walls, pillars, and stairways are lined with six inches of dirt and sawdust, held by two inches of lumber.

5. Man is also told that there are no booby traps, collapsible stairs, or blood spray; this tends to cut down his nervousness.

6. A buzzer system and red lights are at each entrance to provide safety precautions before and during use of range.

7. All entrances are guarded while range is in use.

8. At the end of series of three stabbing dummies, knife is taken away from student to prevent accident.

9. Also, there is fire extinguisher in each room of range.

10. There is no place in range where total darkness prevails while instructor is near student.

<u>Instructions Before Man Enters Range</u>

[To___: Are you one of the quick or one of the dead? The Combat Section respectfully requests your presence (at any time in the next three days, between the hours of 8:00 AM and 5:00 PM) to help eliminate the nest of saboteurs now inhabiting the basement of building A5.

This saboteur den, commonly know as the "house of horrors," will take one hour of your valuable time. You will have ample time to whet your "shooting" eye and to make your will.

Are you a fighting man or just a member of the staff and faculty at Camp Ritchie? Appointments may be made by calling Ext. 102. No special clothing is needed. We will furnish guns, knives, and ammunition (also medical attention).

Sincerely, Miss Test-Ticle Twist.]

Each man reads this: You are equipped with a pistol, twenty-four rounds of ammunition, and a fighting knife. Upon these weapons your life depends as you go down into the darkness. Below are twelve of your enemies awaiting you as you make your way along. You will fire at these enemies in bursts of two shots. You will use your knife at appropriate times.

You will fire directly to your front, to your left, or to your right; <u>you will never fire to your rear</u>. A coach will follow immediately behind you to act as your guide and confessor.

Are you one of the quick or one of the dead?

There are no booby traps, collapsible stairs, or blood baths in the darkness below. If you come out alive, please tell no one else the details of what you have been through.

A REPORT ON THE TEST RELATIVE TO DETERMINING THE MERITS OF VARIOUS TYPES OF HANDGUNS WHEN USED BY THE INSTINCTIVE POINTING TYPE OF SHOOTING
[Document 6d; 4 March 1944]

During the past two years, this officer has been greatly interested in the study and development of the instinctive pointing type of combat handgun shooting and has had an opportunity to train a great many men successfully in this technique. Though a man can be trained by the instinctive pointing method so that he could shoot any handgun successfully in combat without the aid of sights, it became evident over a long period of observation that certain structural differences existing between weapons of different makes (pertaining to design, grips, overall balance and pointing qualities) made a great deal of difference in the initial aptitude and confidence of the individual shooter and his subsequent general combat effectiveness with handguns.

In other words, there are certain weapons which, when picked up by the soldier and fired for the first time, instill in him a feeling of confidence which directly affects his progress during training and his combat effectiveness, particularly when shooting under physical and mental tension.

In an effort to determine the most suitable available handgun for use by soldiers using the instinctive-pointing type of combat firing, the following test was undertaken. It was recognized that in the armed services particularly, the hand must fit the gun and not the gun the hand. It was thought desirable to test available government-issue weapons of different makes and models to determine whether one particular make or type was more suitable for the average soldier than the other.

All firing in this test was done [using] the instinctive pointing method, which demands a combat grip (hard, convulsive), a firing position of a forward aggressive crouch, and the firing of a burst of two shots the instant the weapon is raised from the downward ready position to the eye level.

The comparable weapons involved in this test were fired by the shooter in three groups of two each.

Group I

1. The Colt .38 Detective Special (2-inch barrel) revolver, serial number 479690, blue finish, commercial model and manufacture, checkered wooden grip. This weapon is issued to personnel of the U.S. Army.

2. The Smith & Wesson .38 Special (2-inch barrel) revolver, serial number 0422, sandblast finish, uncheckered wooden grips. It is believed that this weapon is a wartime production weapon only and is the type that is being manufactured and issued to personnel of the U.S. Navy and Coast Guard.

Group II

1. The Colt Commando .38 Special (4-inch barrel) revolver, serial number 37975, sandblast finish, wartime production, plastic checkered grip. This weapon is issued to personnel of the U.S. Army.

2. The Smith & Wesson .38 Special (4-inch barrel) revolver, serial number 540430, sandblast finish, wartime production, uncheckered wooden grip. It is believed that this weapon is also manufactured and issued to personnel of the U.S. Navy and Coast Guard.

Group III

1. The .45 caliber pistol, Model 1911A1, serial number 1860444, manufactured by the Ithaca Gun Company, plastic checkered grip.

2. The Browning 9mm automatic pistol, serial number IT5109, manufactured in Canada for the Chinese government, plastic checkered grip.

These test weapons were picked indiscriminately out of a number available. No alterations or changes of any type were made prior to, during, or after this test. The same weapons were used throughout by all shooters. Twelve rounds (in bursts of two) were fired from each weapon by each man.

Ammunition used was Remington Kleenbore .38 Special Police Service, 158 grain lead bullets. Issue (steel case) ammunition was used in the .45 [FMJ] pistol and 9mm [FMJ] ammunition manufactured in Canada was used in the Browning. The 100 officers and enlisted men who fired this test were picked without warning and had no previous knowledge or introduction to the test. They were merely ordered to appear on the range at such and such an hour and when they arrived, they were oriented and fired. They did not come from any particular organization or group but were picked from 14 different organizations at this center. They [were] members of all branches of service and as indicated, there were all degrees of familiarity with handguns amongst the shooters involved.

Each shooter before firing was told the purpose of the test and was asked to select one weapon out of each group of two which he would prefer to train with and take into combat with him, provided he were given a choice. He was also instructed to make no verbal comment before, during, or after firing while in the presence of other shooters.

After each group firing, [each shooter] was instructed to check in the appropriate column his preference and write a few words as to why it was his preference.

Brand names were not used in distinguishing one weapon from another except in those obvious cases where either the army issued weapon was concerned or a shooter indicated previous familiarity by calling them by the manufacturer's name. Tape of a distinctive color was used on each weapon to enable the individual shooter to clearly identify the weapon of his choice.

Three officers conducting the test made no comment at the time regarding one weapon or another and did not in any way influence the decision or opinion of a shooter in writing down his conclusions.

The following points were checked and noted by the three officers who conducted this test.

1. Whether or not the grip had to be readjusted after each burst of two shots.

2. The spacing between individual shots of the two-shot burst.

3. The size of the shooter's hand.

4. The degree of previous familiarity with handguns in general.

Inasmuch as individual accuracy of the weapon or shooter was not being tested at this time, and because the purpose of the test was to determine preferences in grip, pointing qualities, and "feel," the aiming point for each shooter was a white stake set in a dirt bank at eye level at a distance of 15 feet. The use of the dirt bank also enabled a good means of observation of the difference in spacing between the individual shots of the two-shot burst required from each shooter.

<u>Classification of Shooters</u>

The shooters [involved in] these tests were classified by means of known ability and careful interview by the three conducting officers.

"Expert" classification was interpreted to mean a man who had at least a year's experience as an instructor in the use of handguns and one who was particularly well versed in both the target and combat firing techniques.

"Familiar" was interpreted to be a shooter who had a fair degree of experience with handguns prior to entering the armed forces.

"Average" was interpreted to mean a shooter who had received all of his handgun training in the army and has had only the military instruction and qualification courses.

"Unfamiliar" was interpreted to mean a shooter who had no previous experience or training with handguns whatsoever.

General Results of the Test Firing

1. The Smith & Wesson handguns had the greatest shooter appeal by overwhelming margin, [with] over 75 percent of the shooters selecting this make of weapon.

2. The 9mm Browning automatic was an overwhelming favorite over the .45 pistol, [with] over 79 percent of shooters favoring this weapon.

Analysis of the Reasons for Shooter Preference

The Smith & Wesson revolvers used in the test had two great points of superiority over the comparable Colt models. These were:

1. The wooden grips on the Smith & Wesson are much more hand filling, being longer, broader, and larger in diameter than the ones on the Colt model. I believe also that if the smooth wooden grips on the Smith & Wesson had been checkered, as they were on the comparable Colt, some of the shooters that stated they preferred the Colt would have indicated a preference for the Smith & Wesson, even though it was explained that one being checkered and the other not should not enter into individual choice of respective weapons, inasmuch as the Smith & Wesson could have been obtained with checkered grips if more time had been available.

2. The general structure of the Smith & Wesson frame with respect to the butt is, of course, slightly different than that of the Colt, the principal difference being the more pronounced tang which fits into the "V" between the thumb and forefinger of the shooting hand. This metal tang, which projects higher and is more evident on the Smith & Wesson than on the Colt, enabled a better combat grip, principally because when the weapon was held under a convulsive grip and fired in a burst of two, it prevented the hand from sliding up after the first shot. This resulted in a more even, closely spaced burst of two

shots in the case of the Smith & Wesson, whereas in the Colt, due to the lack of this tang, it was noticeable that the second shot of the two-shot burst was often widely separated from the impact point of the first shot. This was due to the general design of the Colt, which allowed the hand to slide up on the weapon after the recoil of the first shot even though it was gripped tightly. This was particularly evident in the Colt .38 Detective Special.

<u>Comments</u>

1. The more pronounced tang fitting in the "V" of the shooter's hand is definitely more desirable for combat firing, although this feature has been a source of criticism and has been objectionable to a good many target shooters, the reason being, of course, that with [the] relaxed grip necessary in bull's-eye shooting, the recoil of the piece forced the "tang" back into the "V" of the hand and it was not possible to do too much target shooting without a sore spot developing between the thumb and forefinger. It is understood that the Magna grips (developed by Smith & Wesson) were intended to eliminate this objectionable feature insofar as the target shooter was concerned.

2. It was also noted that the grips of the Smith & Wesson revolvers tested seemed to be larger and of different construction than those on their pre-war commercial models. Whether or not these grips are a result of wartime demands and were not made before the war by Smith & Wesson is not known, but they are a definite improvement over the regular commercial grips, which were generally much smaller.

3. Significant is the reaction of the first nine men classed as "experts" on the enclosed chart [not reproduced]. These men have been instructing in combat firing for over one year and were used to, and had only used, the Colt weapons. Irrespective of this, when subjected to the test under the same conditions as the other shooters, every one of them preferred the Smith & Wesson to the comparable Colt, even though most of them had previously enjoyed familiarity with the Colt weapon and little with the Smith & Wesson.

4. It should also be noted in this analysis that the favorable reaction of some shooters to the Colt was because they had very small hands, thus preferring the smaller grip on the Colt, a natural reaction. However, for the vast majority the Smith & Wesson grips were much more satisfactory.

5. The Colt Detective Special is definitely not as

satisfactory a combat weapon as the comparable Smith & Wesson. I would not recommend it for anyone to carry if the Smith & Wesson were available. The Colt is approximately 6 ounces lighter than the comparable Smith & Wesson and this difference in weight, plus the difference in grips and frame structure previously discussed, makes it extremely hard to control when fired double action.

In fact, in our training program here we have had to advocate the use of the little finger underneath the butt to prevent the tendency of the hand to slide up on the weapon when fired. This will almost always be necessary when more than one shot is fired from the Colt Detective Special (as it usually is in combat).

6. The 4-inch Smith & Wesson tested is approximately 3 ounces lighter than the comparable Colt model, yet due to grips, structure, etc., the Smith & Wesson was still preferred over the Colt. This is significant because ordinarily, the heavier the weapon the better it can be controlled, yet in this case due to certain superior qualities of the Smith & Wesson, the lighter weapon was still favored because it could be controlled easier by the majority of the shooters.

7. Smith & Wesson revolvers have a much larger trigger guard than the comparable Colts. This is particularly important if quick draw is emphasized, as the weapon can be drawn and fired with more ease. It was also noted that the edge of the frame of the Colt above the trigger is very sharp in contrast with the more rounded Smith & Wesson shoulder. Because of this sharp shoulder on the Colt and the small trigger guard, men with thick fingers have more difficulty in using it in quick draw, and the sharp shoulder on the Colt hurts the trigger finger when the weapon is gripped tightly and used in double action.

8. Many of the shooters tested stated that the Smith & Wesson had a smoother action. Aside from the mechanical reason (pro and con) I believe this was due to the fact that on the Smith & Wesson revolvers the trigger pull exerted by the trigger finger is more direct, whereas on the Colt the trigger is pulled at more of an angle necessitating more pressure. On the Colt this is noticeable even on the first shot. When firing in bursts of more than one shot on double action, this becomes more noticeable than ever because the recoil after the first shot usually changes the grip on the weapon, making the trigger more difficult to pull because of the increased angle of the trigger finger to it caused by the hand sliding up on the grip.

9. After the conclusion of the test two types of grips were secured for the Smith & Wesson models. One was the standard grip, checkered in place of plain (as were the weapons used in the test), the other was the Magna type grip. By having a number of shooters who had indicated a preference for the Smith & Wesson in the original test fire of the Smith & Wesson weapons with both types of grips, the following general conclusions were reached:

a. That the checkered grip in both the Magna and standard grips was much more preferable than the plain grip which is now furnished on the government issue Smith & Wesson weapons.

b. That the checkered wooden grip was much superior to the plastic checkered grip such as is found on the Colt Commando, the plastic checkered grip being more susceptible to slipping under combat conditions such as sweaty hands, etc., than the wooden checkered grip.

c. Generally, the standard grip produced just as good results in combat firing as the Magna type. Although most shooters preferred the feel of the Magna grips, I do not believe that they add greatly to accuracy and pointing qualities over the standard grip now furnished on the navy issue weapons. However, the Magna grip does eliminate the possibility always present in the Smith & Wesson (especially under combat tension) of the thumb pushing forward the cylinder release thereby preventing the weapon from firing.

10. A subsequent test of the Colt Banker's Special (commercial model) against a Smith & Wesson .38 caliber, 2-inch barrel (cut-down British model) led me to the following conclusions:

a. That the objectionable grip and frame features present in the Colt construction still apply to weapons firing the .38 Smith & Wesson cartridge. However, of the two Colt weapons (the Banker's Special and the Detective Special), the Banker's Special is definitely a better combat weapon when fired double action because it can be handled, fired, and controlled (due to its lighter cartridge) in a much better manner than the Detective Special, which fires a cartridge entirely too powerful for its weight, design, and structure.

The cut-down weapon (2-inch barrel British model .38 Smith & Wesson cartridge) just received would, in my opinion, be a very desirable weapon commercially in competition with the Banker's Special, but I believe it would be necessary to lighten the frame considerably to make it approximate more closely the weight of the comparable Colt model (Banker's Special). The grip and frame structure, however, should be retained. As a military weapon, it would be very satisfactory just as received, but the structural superiority of the Smith & Wesson would enable a considerable lightening of this weapon for commercial purposes without any loss in pointing and handling qualities.

The 9mm Browning Automatic

The 9mm Browning was a decided favorite over the .45 automatic pistol. Obviously the difference in caliber between 9mm (.35 caliber) and .45 caliber is quite great. This is the reason, we are told, that the U.S. Army adopted the .45 caliber, although every other army in the world seems to be more satisfied with the cartridge of the 9mm type for use in its handguns and submachine guns. I personally contend that hits are what count and if you can take one weapon and make hits better with it, although it is of lighter caliber than one [that uses] heavier cartridges (which is more difficult to shoot), it is advisable for the average man to adopt the lighter weapon as long as the caliber of this weapon is not too small.

It should also be considered that the Canadian weapon holds twice as many rounds as the .45 pistol and that it is still approximately 4 ounces lighter with a full loaded magazine than the loaded .45 automatic pistol.

The .45 automatic has been the subject of a great deal of criticism throughout all branches of our armed forces, and it has long been evident that military shooters have very little confidence in the weapon for target firing, let alone for use in combat. The general structure of the pistol, which I understand is from an earlier Browning patent, has made it notorious for its poor pointing qualities. This is particularly so when it is used under combat conditions necessitating an extremely tight grip (barrel pointing down). The combination of the tight grip and shoving the weapon toward the target when forced to fire without using the sights has resulted in too much inaccuracy and a consequent lack of confidence in the weapon in the hands of the average shooter. This condition will always be present as long as the training for combat is done by the standard target method now employed by the army. Good accuracy and confidence in the weapon can be developed, however, by using the instinctive pointing method. The unpopularity of the .45 pistol and the poor results obtained by the average shooter is attributed to be the cause for the issuance of the [M1] carbine.

The general design of the [9mm Browning automatic] seems to have eliminated to a great extent the objectionable structural features of the .45 pistol. The staggered type magazine, enabling the weapon to hold 13 rounds, has also resulted in a fatter grip on the Browning, which makes it more hand filling and consequently easier to point, control, and fire. This also makes the difference in recoil between it and its competitor, the .45 pistol, much more noticeable, although it is recognized that the recoil is much less in the 9mm because of the difference in caliber of the two weapons.

The grip on the Browning is also more satisfactory because the indentation in the back of the grip underneath the slide is [in] approximately the same spot as the safety release feature found on the .45 pistol.

If I were given a chance to choose an automatic weapon with which I would train troops and [which I would] want them to use in combat, I would pick the Browning because of the above-mentioned reasons. However, it should be borne in mind that outstanding results can be and are being obtained with the .45 pistol by making the instinctive pointing type of approach to its use in training and combat. Even if the present target type of army instruction in the use of the handgun were to be continued (factors of production and supply not influencing the choice), the average shooter would do even better in the target type training with the Browning than he does with the .45 caliber pistol, because its general characteristics are much more desirable for the aimed type of shooting as well as for the instinctive pointing type.

CHAPTER SIX

MITC: GUTTER FIGHTING

KNIFE FIGHTING
[Document 7a; 19 March 1943]

1. Introduction

Every American soldier, officer or enlisted man, should be issued a fighting knife and trained in its use. Although in World War I our men were issued a trench knife for close-quarter work, to date in this conflict our soldiers have not been issued a true fighting knife, the deadliest of all close-quarter weapons.

In the present conflict the fighting knife has two major uses, one as a reserve weapon when all others fail, and the other for specific reasons, such as assassinations, sentry killing, or in any case where silence and quick killing efficiency is desired.

That it is important as a major weapon has lately been evident from reactions and reports from the Pacific theater, where our enemies have put it to such good use. In the European theater the commando type troops have used it with success, and in most of the armies, both Allied and Axis, some sort of knife has been adopted and issued to all military personnel, although little definite instruction in its use seems to have been given the troops carrying it.

In certain areas it has played an important part in hand-to-hand combat. Yugoslavs, Greeks, and other natives of the Balkan area, the Finns, and some Russian outfits have made good use of the fighting knife.

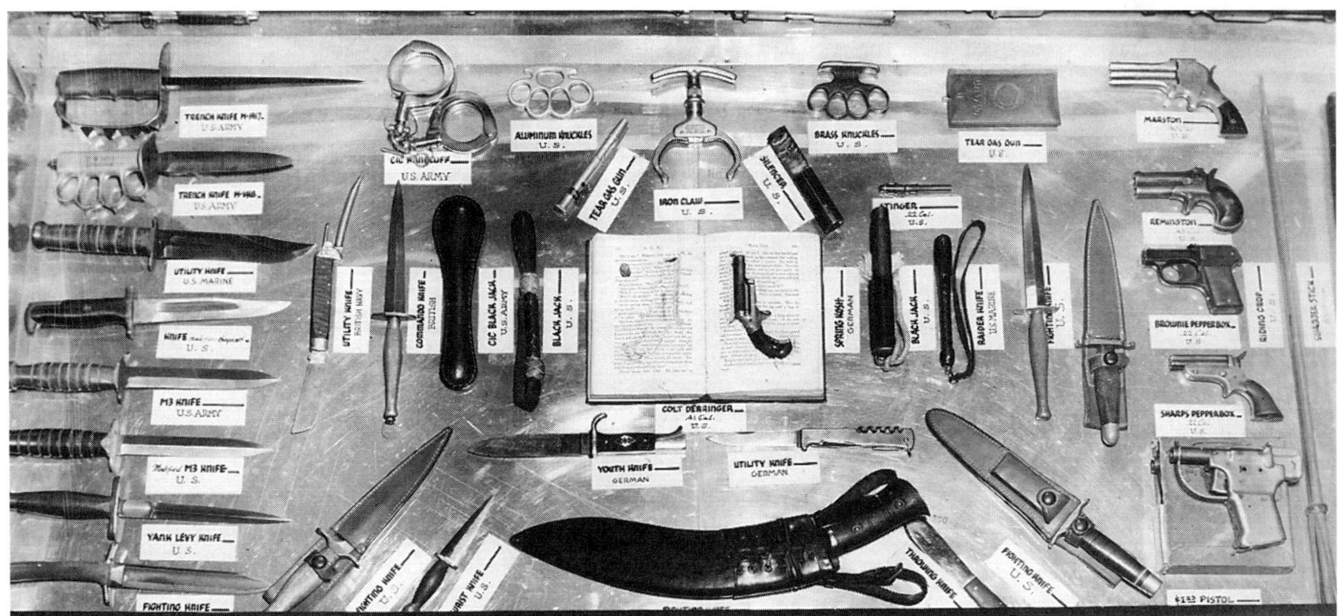

MITC weapons display that included a variety of edged and blunt trauma items that could be encountered or used overseas.

<u>Knife Throwing:</u> Before going into actual knife fighting technique, we should discount knife throwing as a practical method of combat. There are few individuals in the world who can pick up a knife and throw it at a moving object at an unknown distance and hit a vital spot.

In the main, knife throwing is an art relegated to vaudeville and stage. The reason for this is that to throw a knife properly, the exact distance from the thrower to the target must be known, because the knife turns end over end as it travels through the air. The thrower, therefore, must know this distance to be able to control the number of turns the knife makes, so that it may hit the target point first.

There are some methods of knife throwing at close ranges without the blade's turning over in the air, but considering the movement of the target, heavy clothing and the fact that if you miss, you are without a weapon, it is easy to see that knife throwing is not too practical.

<u>2. Psychological Effects</u>

There are definite psychological considerations in regard to knife fighting, which pertain to both the user and the enemy. In the first place, unless the knife is considered a personal weapon by the individual, such as is common among certain groups, the untrained will have a noticeable aversion to thinking of the knife as a weapon to use in combat.

This is especially true of the ordinary American soldier who would much rather use his fists in close-quarter fighting than a knife, because, generally speaking, the knife is little used as a weapon in civilian life. This is a very good reason why it is important to train our personnel in using a knife.

This psychological barrier must be overcome, and the soldier must achieve skill in handling the knife as a weapon. The average American doughboy, when shown a fighting knife the first time, will have an aversion to its use as a fighting implement.

This same feeling is apparent in preliminary stages of bayonet training. However, once the infantryman has run the bayonet course and used the bayonet on dummies, the killing instinct becomes aroused to the point where he has confidence in the weapon and is not averse to using it.

The same thing applies to knife training and the same result will be obtained if the individuals are taught to use the knife properly and if dummies, which can be slashed and cut, are used in the course of instruction.

An excellent example of the psychological effect [of the knife] on enemy troops occurred during the early days of the Libyan campaign against the Italians. [Gurkha] troops on the allied side were particularly skilled in using the knife. They were also excellent stalkers.

It was the practice along a certain sector for these [Gurkhas] to crawl into the ranks of sleeping Italians where the knife was used to slit the throat of one of the group only; upon awakening, the soldiers, seeing a dead comrade with his throat slit, would be extremely shaken. This contributed to a general lowering of the Italian morale, and in the long run contributed a great deal to their surrender.

To the untrained man, the appearance of a knife in the hand of an enemy causes panic. This is heightened by the use of a bright, flashing blade in place of a blade of blue steel. There is a definite advantage to the attacker who uses a bright blade instead of a darkened one.

Although less conspicuous, the knife with the darkened blade is in reality not much advantage, because as the color of the blade is produced by a bluing process, it will in a very short time (due to sharpening, wear in the sheath, etc.) wear off, leaving the blade bright.

<u>3. Essential Qualities of a Good Fighting Knife</u>

Knives at present fall into two general categories: those designed for straight fighting and the multi-purpose weapon, called the utility knife. Others have been issued for fighting alone, but their general design has been poor.

Actually, the best fighting knife should be constructed with a stainless steel blade and a dark handle, which will not wear bright so that light reflection when the knife is in the sheath is not possible.

As the knife can be for either cutting or thrusting, the ideal weapon for close-in fighting combines both qualities, plus extreme maneuverability. This last feature is very important because, due to the handle design, which should be like that of a fencing foil, it can be used for cutting and thrusting in any direction whatever without changing the grip.

The weight in such a fighting knife is toward the hilt. The blade is about 6 inches in length, is double-edged and tapers to a point. This length blade is ideal for balance, is good for both the cut and thrust, and is long enough to penetrate heavy clothing without losing its effectiveness.

Its width at its widest part near the guard should not be over 1 inch. It can be either hollow ground or tapered evenly toward both edges from the strengthening ridge which runs down the center of the blade until it reaches the point of the knife.

The handle construction is round or oval in shape and is such that the largest diameter is toward the center and tapers off toward the guard as well as the butt. The overall weight is approximately 10 ounces. The handle, in addition to being rounded, is checkered.

Such a knife, with balance toward the handle, lends itself more easily to maneuverability, is more easily passed from hand to hand, and with weight in the handle, gives a better grip for passing, thrusting, and slashing. Its very design makes it a true fighting knife, combining both slashing and cutting qualities due to its double edge. The double edge is also desirable in preventing an opponent from wresting it from the hand of the user. The opponent cannot grasp its blade without a severe cut.

4. Technique of Handling

A simple demonstration of a knife gripped as in the fencing foil and a knife gripped first with the wrist on the upper side and then on the lower side, showing limitations, clearly presents the advantage of the true fighting knife over the utility knife. This advantage, however, needs more than actual demonstration to take effect, and if you as an individual were forced to use a knife in the course of a few hours after the demonstration you would probably grip it in an unskilled manner, allowing only a downward or an upward thrust. This can be attributed to lack of practice.

The proper grip on the handle of a knife of this type is as follows: it lies across the palm of the hand diagonally. The small part of the handle next to the cross guard is grasped by the thumb and forefinger. The middle finger is place across the handle at the point where its largest diameter occurs.

With the knife held in this fashion, it is very easy to maneuver it in all directions, by controlling the direction of the blade by the combined movement of the fore and middle fingers plus a turning of the wrist. When the palm is turned up it is possible (holding the knife in the right hand) to slash to the [left].

When the palm is turned down it is possible to slash to the [right]. The thrust can be executed from either the palm up or down position.

At the time of contact in the thrust or slash, the knife is grasped tightly by all fingers. The initial controlling grip of the fore and middle fingers has not changed and the blade becomes a mere continuation of the arm.

Such knife manipulation is easy and can be acquired after a few hours' practice, but only if the handle is generally constructed after the lines described above. The handle herein described is round. However, the handle of the same general proportion in the oval shape works equally well with practice. After the student has been shown the vulnerable spots, he should take a real knife and practice its manipulation facing a dummy. This dummy can be an old pair of overalls filled with straw or any other suitable replica of a man's body which has arms and legs.

Practice slowly at first, executing thrusts and slashes always from the crouch; speed up the tempo as the practice goes along and select spots to hit as you practice. About 6 hours of such practice will give the student an extreme amount of confidence in his weapon and a skill in its use which will place him well above the average knife wielder.

In those cases where you are in the open and the opponent can see you, attack from a crouch with the left hand forward and the knife held with the handle across the palm of the right hand close to the body.

The left hand will act as a guard and a foil or parry, which will help in getting an opening for a slash or thrust. When the man is in the crouch with his left hand forward to parry, he is in a position of extreme mobility, due to the flexing of his knees and to the fact that he is in perfect balance. In the crouch he is also protecting his vital midsection and throat area from possible vital thrusts from his opponent who might likewise be armed with a knife. He is also in a position where he can possibly foil the usual knife defenses if his opponent is unarmed, such as with a chair, a club or any other object which may be used to strike or throw.

At this point dummy knives, constructed particularly in the handle as a fighting knife, are placed in the hands of the students. Students practice thrusts and slashes.

Vital Spots: A man when attacked from the front with a blade has two spots which he instinctively protects. They are the throat and the stomach (the abdominal section). Perhaps the reason he instinctively protects these two areas is that they are easy to reach, but in any event the psychological effect of a knife wound in these areas, regardless of whether it is serious, is so great that the victim is usually momentarily incapacitated.

The throat area is susceptible to either the thrust or the slash, the thrust being more effective when driven in the hollow of the base of the throat just below the Adam's apple. A thrust there into the jugular vein or a slash on either side of the neck, cutting the arteries which furnish the blood to the brain, results in extreme loss of blood and death in a very short time.

Thrusts in the abdominal area, which can be combined with the slash as the knife is withdrawn, have a great shocking effect upon the individual and usually incapacitate him to the point where another blow can be given with the weapon before he has a chance to recover. A deep wound in the abdominal area will cause death if unattended, but is much slower than a good thrust or slash in the throat area.

The heart is, of course, a vital spot for a thrust, but the protection of the ribs makes it difficult to hit. In some cases knife thrusts toward the heart have been stopped by the ribs and the point of the knife broken off by the bony structure without causing a vital wound. Usually, however, the blade will slide off the rib and go into the vital area. The heart thrust is, of course, immediately fatal.

It is possible to get an effective slash across the sides of the throat from the rear, but one of the most effective knife blows in the rear area of the victim is that delivered into the kidney or small of the back area. Penetration here in the form of a deep thrust will cause a great shock and internal hemorrhage, but not necessarily death. This back or kidney thrust is best used in sentry attack, as will be explained later.

The vital areas are still the heart, throat and abdominal sections, and all knife thrusts or slashes should be preliminary to the vital killing stroke delivered into these areas.

The slash can be effectively used to sever the tendons on the inside of the wrists, which is most effective against a person who is trying to protect himself from the knife and has his arms outstretched to do so. The slash renders his hand useless. A slash across the large muscle of the biceps has the same effect.

A slash on the inside of the thigh or arm will cut arteries, and will also incapacitate if delivered deep enough. The slashes of these areas, in addition to disabling the opponent, cut various veins and arteries and, if left unattended, will cause death from loss of blood.

Carrying the Knife: To cover the various places in which knives are carried, let us say first that the knife should be carried in a place where the bearer can, with the least possible effort and with the most speed, draw it from its sheath. This place where he carries the weapon may vary greatly due to local customs and the type of garment the carrier wears.

Knives have been carried successfully in the following places: in a sheath on the side, down the back of the neck, up the sleeve, stuck in the top of a boot or legging, in a sheath sewn inside a front pocket, under the lapel of a suit coat, in the crown of a hat, between the belt and the trousers, strapped inside the thigh beneath the trouser leg, in a sheath sewn diagonally across the chest, on a vest and in any other place combining both concealment and the element of surprise to the advantage of the user.

As in the carrying of small arms, there is no one best place to carry a knife. Each individual has his own ideas, but once a place has been decided upon, let the knife user carry it there constantly and practice its draw from that location, so that he will be able to use it with the greatest speed and with as much instinctive movement as possible.

5. Sentry Killing and Assassination

In sentry killing, all things regarding the approach and initial attack described for use with a Japanese strangle apply [see Chapter Seven]. The approach from the rear is naturally a noiseless one. At the time of rising a few feet in the rear of the victim, the knife should either be taken from the sheath, where it has been during the crawl, or taken from the teeth where it might have been carried. The attack is launched from a distance of not less than 5 feet from the victim and is initiated as soon as the attacker has arrived at that spot.

This immediate attack is important because of the animal instinct, emphasized by keeping your eyes steadily on the victim as you approach, which will warn him that someone is approaching and watching him. The thrust of the knife into the middle of his back or the left or right kidney section is executed at the end of the leap to attack. At the same time the free hand is clasped over the mouth and the nose of the victim, pulling him backwards off balance.

The thrust into the kidney area has initially a great shocking effect, and no outcry will occur if the free hand goes over the nose and mouth at the time of the thrust. Pull the victim back on the blade from behind and, maintaining the same grip on nose and mouth, lift the head up and slash the jugular vein.

One method of using the knife in assassination is worthy of mention because it is as old as history and is practiced throughout the occupied countries today. Fortunately members of the Gestapo and the local [collaborators] have been the majority of the victims.

The assassin spies his victims in a crowd and approaches him from the front. His knife is held in his hand with the hilt down and the blade lying flat along the inside of his forearm or concealed up his sleeve. The handle is, of course, concealed by his fingers.

The assassin with the knife in this position passes the intended victim walking toward him, and as he reaches a point directly opposite the victim, a simple movement of the wrist frees the blade and a short arm movement plunges it into the kidney area of the victim. The knife is either left sticking in the wound or is pulled out and the assassin walks on through the crowd, his movement generally undetected.

6. Defense Against the Knife

In a majority of cases where an individual comes up against a knife in the hands of an enemy, he will never get a chance to see it coming until it is too late to do anything about it. This condition is due to poor light or weather conditions during which the use of such bladed weapons is ideal, as well as the fact that the knife and its draw from the sheath are usually well concealed.

Keep Knife-Man Away From Body: At extremely close quarters there is not time enough to make any defense against a knife other than an instinctive arm parry or block. Under circumstances where you are able to see the attack coming, there are certain defenses which are very effective, if you have time to execute them. If possible, never let a man with a knife get within striking distance.

Throw your empty weapons, your helmet, or a handful of dirt into the attacker's eyes, or pick up a club or other weapon readily at hand. If these means of defense do not stop him or are not available, there are defenses at close quarters by which you can protect yourself from the blade and after doing so execute an attack.

The following paragraphs discuss defense possibilities of dealing with a knife in the right hand of the enemy, provided you, as recipient of the attack, are also right-handed. By the same token, the left-handed individual, by reversing the described methods, can obtain the same results.

Force of Knife Blow: Any knife or other type of defense should depend as much as possible upon an individual's instinctive reaction. In other words, it is much easier and surer for a right-handed individual to use the right hand to ward off or parry a blow than it would be for him to use his left hand in the initial phase of the defense.

When the knife is raised above the individual's head in a position to make a downward thrust, the best time to start the defense is before momentum and power are applied in the downward, sweeping motion. In most cases, however, the hand grasping the knife will be coming downward toward the victim with all the power and force the attacker can command to satisfy his lust to kill.

Fallacy of the Common Block: Most common knife defenses shown involve a "block" by grasping the wrist or by using the forearm. These do not take into account the extreme force of the downward thrust and its resultant impact, which will cause the blade to crash through these defenses to penetrate a vital body area.

The next common knife defense advocated is the one which utilizes a grip of the wrist by the left hand followed by an arm lock, or by a block of the downward thrust, using the right forearm. In these two cases the momentum of the knife will ordinarily cause a crash through to the knife's objective.

Another danger of using the common defense of grasping the "knife wrist" with the left hand is the fact that such a method depends upon good light and perfect timing to enable a grip on the moving knife wrist. If the blow is sweeping down with great force and the recipient tries to grip the wrist in his left hand, it may be such that the thumb side of the gripping hand is liable to give way, thus allowing the thrust to continue toward its goal.

Another disadvantage of this common type of defense is that the gripping initial movement, which is usually instinctive, is with the left, not the master hand. Therefore a great amount of practice is necessary before such a grip can become instinctive.

Use of Master Hand: The best grip defense against the downward sweep of a knife, therefore, is one which utilizes the master hand, which is the right hand in the case of most individuals, and takes the force on the palm and finger side. By gripping the "knife wrist" with the right hand, you also pivot your body as you reach forward, thus protecting exposed vital body areas.

From the position of the knife wrist grasped in the right hand, you may apply an arm lock or any of the other means of attack which have been mentioned before, the use of which can be determined best after some practice. The wrist grasped with the right hand is also a risky proposition in poor light, but inasmuch as the body pivots in its execution, there is less likelihood of receiving an incapacitating wound.

The Parry: A more certain defense against the downward knife thrust is the parry, diverting the power of the thrust as it sweeps downward. This is better because the whole length of the arm can be used. By using the right arm parry, a thrust to the right will cause the hand holding the knife to follow down along the outside of the body. Even if the parry is not entirely successful, a flesh wound in a non-vital area will be the result.

Here again, the recipient of the attack takes advantage of the instinctive movement to thrust his master hand above his head in order to protect himself from the downward blow, the only difference being that the movement of the right arm is a sweep to the right in place of a block. Conversely, one can parry the downward blow of the right-handed man by using the left arm to parry to the outside; but the chance here of the knife crashing through, if the parry were unsuccessful, is greater, because the body is directly facing the knife man, whereas when you use the right arm the trunk of the body is turned.

When you are faced with a knife held in the hands of an enemy for upward thrust into your abdominal region, the parry is again the best means of defense. The parry can be executed either with the right or left arm as follows: As the attacker makes an upward thrust, sweep your right arm across the front of your body and catch the upward moving knife arm on the outside of your arm. This will cause the direction of the thrust to be diverted to your right, or outside of your body. The left arm may be used to take the initial impact to parry the weapon to the right also, but better timing is necessary if the left is used.

Any sweep of the arm in any direction, such as in a parry, causes the body to pivot naturally out of the line of the thrust.

Use of Feet: One of the most effective means of defending oneself against a knife man advancing to attack in a crouch position with the knife held close to his body is to use the side of your foot to kick the knee of the advanced leg of the attacker.

If you find yourself surrounded in a room where individuals would be likely to use knives, back into a corner and use your feet to keep them out of arm's reach.

Use of Chair: The "chair defense" against a knife man is good, provided you have a chair handy. Grip it by the back and point the legs at your attacker. Advance toward him, making short jabs as you advance.

The principle involved here is the same as that used in lion taming. The knife man cannot watch all four legs of the chair at once. Consequently he becomes confused and is more susceptible to blows from your feet which can be directed toward his body in conjunction with a thrust of the chair.

Use of Stick: Another defense, which may be employed against a skilled knife fighter who [holds the knife with] the hilt lying across the palm of the hand, is the use of a small stick. The stick is grasped in the right hand in such a manner that the length runs down the inside of the forearm. After some practice you can make it virtually impossible for the knife man to reach into your vital body areas. Use the forearm to parry any attempt to thrust. The stick lies along the inside of the forearm and wrist in such a manner as to protect the tendons and arteries against a disabling cut or slash. A cut on the outside, or bony part of the forearm will have little "ham-stringing" effect.

As mentioned previously, these defenses are only possible and practicable after you have had ample time to see the knife coming toward you. The best stick defense is the one which involves the use of a limb or club the size of a baseball bat, using it with both hands in the traditional manner.

Value of Surprise: Remember that the element of surprise is very great against the knife man if you can take the offensive in conjunction with the parry or block.

As to any definite means or methods to finish him off during the attack, that depends upon you, and what you can do best according to what you have gained through practice of the above mentioned defense tactics.

7. Conclusion

Recent reports from both enlisted men and officers of army and Marine forces engaged in the Pacific theater of war have all stressed the desirability of a knife for troops. The requests have varied from the request for the utility weapon, which admittedly is a necessity in the heavy jungle growth, to requests for a pure fighting knife.

When early Marine units departed for the Pacific theater last year, the commanders who had considerable experience in jungle warfare in [Central] and South America requisitioned from hardware stores and other sources of supply on the Pacific coast large clasp knives, hunting knives and other knives available which would be suitable for jungle work.

Particularly in the Pacific theater, the knife has been proven to be a very important weapon, because the very nature of jungle warfare makes it close-quarter work, where the bladed weapon is particularly useful, especially in the dark.

As mentioned, American troops should be thoroughly introduced to the knife as a combat weapon and trained in its use. Large numbers of troops will sooner or later come into contact with it in the theaters of operations. The psychological effect on the individual soldier will be much less if he has had preliminary training and instruction in the use of the knife, particularly when he is faced with a shiny blade in the hands of an enemy.

Knife and Knife Defense
[Document 7b; 20 April 1945]

I. Number of Instructors
1. One officer per each two sections of enlisted men.
2. One officer for instruction of each thirty-five officers.
3. One assistant for each officer instructor.

II. Place to be Held
1. Parade Ground.
2. Range (when not in use).

III. Equipment Needed
1. Twelve Knives
a. Commando [knife]
b. M3 [trench knife]
c. Modified M3 [trench knife]
d. Trench knife M1917
e. Biddle fighting knife
f. Marine utility [Ka-Bar]
g. Throwing knife
h. Assassin's knife
i. German knife
j. Our own utility knife [see Document 7c]
k. Trench knife
l. German youth knife

IV. Preparations by Instructor
1. Check with Chief Instructor.
2. Check schedule book.
3. Coordinate movement of classes to area.

V. Knife and Knife Defense

Knife defense to be shown first.

The class should be shown several types of fighting knives and utility knives and the advantages of fighting knives should be shown. However, when they pair off to work on each other, tent pegs should be used in place of knives. Each parry, as well as each method of attack, should be taken up separately. Show proper grip on the knife and vital spots of the body.

Dummies should be used to allow students to get the "feel" of actually using cold steel on an object.

When showing how a trained man uses a knife, point out the vulnerable spots to attack. Also mention the time it takes to induce death by attacking those vulnerable spots.

Applegate demonstrates the guard position for knife fighting as taught at MITC. The lead arm is used to make an opening for the knife hand.

Slash or Stab Artery	Depth to Enter Knife	Unconscious	or Dead In
Radial	1/4″	30 sec.	2 min.
Brachial	1/2″	14 sec.	1 1/2 min.
Carotid	1 1/2″	5 sec.	12 sec.
Subclavian	2 1/2″	2 sec.	3 1/2 sec.

This period will deal with the fighting knife: features and points for consideration in the selection of a fighting knife, the technique of using a knife, and defense against knife attack.

A true fighting knife must be considered from the standpoint of its use as a fighting knife alone; actually there is no happy medium between a fighting knife and a utility knife. A knife cannot be designed that will meet the final requirements of both. The utility knife of the design of the commercial hunting knife has been used as a fighting knife on occasion, but there are several features of such a knife that detract from its value as a fighting knife. For example, the utility knife has only one cutting edge, limiting the user to only one type of slashing stroke; the blade is too wide at the point to be an effective thrusting weapon; the handle is large and shaped to fit the hand for a conventional overhand or underhand grip. This necessitates a straight upward or downward thrust. The knife is blade heavy. This, in combination with the shape of the handle, makes it unwieldy to manipulate (display utility knife).

The true fighting knife on the other hand should have the features of a knife adapted to fighting and nothing else. It should have a keen double edge for two-directional slashing, a narrow sharp point for easy penetration in a thrust, a round checkered handle tapering to approximately half [the] center diameter at the butt, the point of balance being slightly toward the handle. These features of the handle lend to greater maneuverability and flexibility in the use of the knife. Suggested overall dimensions: handle—13/16 inch at largest diameter, tapering to 7/16 inch at butt; guard—2 x 5/8 inches; blade—1/4 inch thick x 7/8 inch wide at guard, 11/64 inch thick x 11/32 inch at spot 2 inches from point, tapering to a thin

Applegate shows a reversed knife grip used in slashing attacks by some knife fighters.

sharp point; length of blade, 7 inches; length of handle, 4 1/2 inches. Although a blue blade will not flash in light and will therefore be less conspicuous upon exposure, a bright and flashing blade is preferred, as the bright blade has more of a threatening appearance and will have a psychologically demoralizing effect on the attacked. As for concealing the conspicuous blade, it is simple enough to keep the blade in a sheath until it is ready to be used. The handle, not being concealed by the sheath, should be of a dull finish that will not reflect light.

In using the fighting knife it should not be held with a fist, thus permitting a common upward or downward thrust; this would not be taking advantage of the maneuverability and slashing possibilities of the weapon. The handle should rather be grasped much in the same manner as a fencing foil is held, with the handle lying diagonally across the palm, the thumb and forefinger opposite each other, next to the guard, and with the middle finger grasping the handle where the largest diameter occurs (demonstrate proper grip).

In the stance for attack with a knife, the attacker should be crouched and in the balanced position, his left hand should be extended for parrying and balance and the knife hand should be close to the side, the knife held with the palm up and ready for commitment. The entire position of the body should be one of balance and crouched readiness (demonstrate stance).

In the actual technique of using the knife, do not draw the knife until it is ready to be used. Attack with the knife in the right hand and a handful of dirt in the left; throw the dirt in the opponent's eyes and follow through immediately with a thrust with the knife to the stomach. The dirt in the eyes will distract his attention from the knife and it may be used without opposition to the thrust. Parts of the body vulnerable to various attacks with

In a position to be avoided, Applegate deflects a knife held in the upward swinging "hammer" grip.

the knife are as follows (demonstrate): thrust to jugular vein; slash to outside of neck; thrust to stomach; slash across biceps and inside of wrists; and slash to inside of thigh. A thrust to the heart is not advisable in the initial attack as the ribs make this vital organ difficult to hit in a quick thrust.

In defense against a knife it is of utmost importance to keep the attacker away from the body; avoid closing with a knife man until it is absolutely unavoidable. Keep him at a distance by throwing your empty weapons, helmet or dirt in his eyes, or defend yourself with a club or some other makeshift weapon readily at hand. If all these means fail, then and only then should knife defense at close quarters be considered.

In defense against the conventional amateur knifer's downward thrust, there are two main methods to develop: the parry of the downward thrust and the checking of the knife arm by means of an armlock before the downward stroke is begun. Attempting to block the downward stroke of a knifer's arm is of doubtful value. A knife man of average strength can force his knife through a common block if his downward stroke has gained momentum. Under good light conditions and upon a favorable opportunity, the defense mentioned before can be effected by anticipating the blow and grasping the knife wrist of the attacker and applying an armlock before the downward thrust has gained momentum.

This may be accomplished by either the right or left hand as the initial check and then following through with a lock by the free hand (demonstrate). In this method, checking with the right hand is always considered more practical as the body will pivot sideways to the attacker, thus protecting vital body areas; also, the use of the right hand is more natural to the average man. As [previously] stated, however, this method of defense must tie in with favorable light conditions and opportunity. It is not considered to be foolproof.

A more reliable defense against a downward thrust is the parry with the arm extended and straight, thus diverting the thrust as it sweeps downward (demonstrate). This parry with the right arm to the right will guide the attacker's arm down the outside of the body and will inflict none or only slight wounds. Parry with the left hand can be done but the use of the right is preferred (demonstrate). This same technique of parrying can be used against an upward thrust toward the abdomen. The parry can be made with the left or right hand in the following manner: on the upward thrust of the attacker's knife, sweep your right arm across the front of your body and intercept the upward moving knife arm on the outside of your right arm. The left arm may be used to divert the upward thrust to the right also, but better timing is required (demonstrate).

In defense against the knife man who is apparently familiar with the techniques of knife fighting (he will manifest this by grip, stance, etc.), it is not advisable to attempt to block or parry an attack. The practical defenses against such an attack are as follows: kick to the attacker's advanced knee (demonstrate), parry with a small stick held on the inside of the forearm (demonstrate), or advance on the attacker with a chair, if available. The chair is held by the back with the legs pointing toward the attacker.

In summary of this period it would be well to stress the following points: choose a fighting knife with an eye to its use as a fighting knife only. It should have a sharp tapering point, a double edge, and a handle that lends itself to maneuverability. In using a fighting knife, grip the handle in such a manner that the knife can be maneuvered for slashing and thrusting, keep a crouched and balanced position, and keep the left hand advanced to parry (demonstrate). Thrust and slash to the aforementioned parts of the body (demonstrate), and attack with a handful of dirt in the left hand. In knife defense the block is not to be attempted; the parry is the only defense of practical value against an upward or downward thrust. Against a professional knife man, kick to the advanced knee, parry with a stick, or shield yourself with a chair (demonstrate).

GENERAL UTILITY FIGHTING KNIFE
[Document 7c; n.d.]
[Report on knife designed and submitted to the Infantry Board]

Although it is understood at this time that the army is contemplating issue to troops of a knife to replace the World War trench knife [M1917 and 1918] this design concerning the enclosed type of knife is submitted with a short explanation as to its use, particularly in theaters such as the Pacific.

The weapon, which in design is not unlike the medieval short sword or the British smatchet, is double-edged and combines both thrusting and extreme slashing or chopping qualities. It is as close to being the all-purpose jungle weapon as this officer can conceive. The weight, which is less than 2 pounds, makes it a good utility weapon for cutting jungle growth and also at close quarters

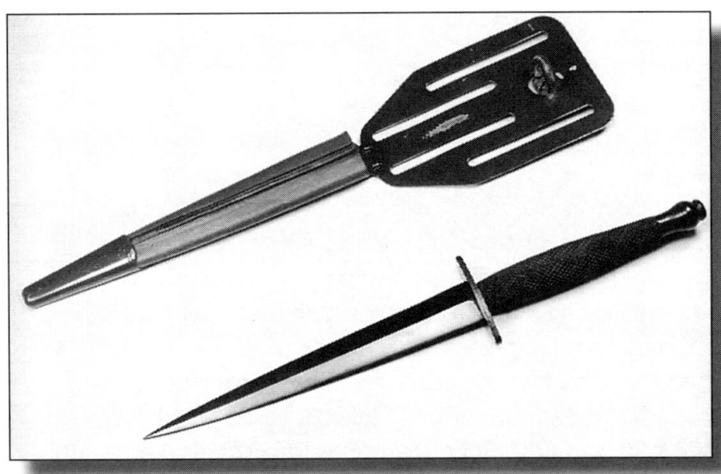

The fighting knife produced for the OSS and based on the Fairbairn and Sykes design. It was considered by Applegate to be the best knife design for the methods of fighting taught at Area B and Camp Ritchie. (Photo courtesy of John W. Brunner.)

it is a deadly fighting instrument. This combination of thrusting and slashing qualities is of extreme importance when considering it for use as a weapon.

The trouble with most utility type knives is the fact that due to their general design, balance, and weight they are awkward and ill-adapted to strokes of the slashing or chopping type. Whenever a knife of the type [Ka-Bar] now being issued to the Marines is used, its very construction limits its use because of the way the handle is gripped. When you grip the handle of a bladed weapon the same way you do when you grip the handle of a small ax, you immediately limit yourself to an upward thrust because your arm pivots on the shoulder; when you grip it on the underside you limit it to a downward thrust. In neither case do you take advantage of the cutting or slicing qualities of the knife.

If the knife is heavy enough, and the balance correct, it can be used as a chopping or slashing weapon, utilizing the cutting qualities of the edge, but in the case of a large hunting knife this is not usually true. The hunting knife type, because of its single edge factor (although it is sharpened about two inches on the top side), does not lend itself to make as easy a thrust as the double edged weapon. If troops are not going to be trained in the use of a knife it is best when issuing it to them to give them a knife of the type which they can do the most with by using it in the natural manner (chopping, slashing, and thrusting).

The enclosed article [not included] on the knife advocates the issue to troops of a fighting knife which has nothing but fighting characteristics plus a weapon for utility work, which naturally could be used as a weapon when circumstances arose. The design enclosed in the rear of the article shows one type of design for a true fighting knife. The enclosed specifications [not included], with picture attached, would be this officer's recommendation for a utility fighting weapon if only one knife were issued. The knife from which these specifications were drawn is available for inspection if the board so desires.

SENTRY KILLING
[Document 8; 20 April 1945]

I. Number of Instructors
1. One officer instructor for each group of two sections of enlisted men.
2. One officer instructor for instruction of officers.
3. One assistant instructor for each officer instructor.
II. Place to be Held
1. Range or parade ground.
III. Equipment Needed
1. Sticks, one per two men.
2. Garrotes, one per two men.
3. Tent pegs, one per two men.
4. One knife, modified M3.
IV. Preparation by Instructor
1. Check with chief instructor.
2. Check schedule book.
3. Coordinate all movements of classes.
V. Sentry Killing (Stick, Garrote, Knife)

The use of the stick, garrote, knife. Explain the proper approach to sentry. The period starts with the class assembled in a semicircle around the instructor. First, a [quick] review will be given on the preceding hour. Then about 10 minutes will be devoted to [strangulation]. Two men will be called forth from the group to demonstrate the Jap strangle [see Part Seven] and two other students will demonstrate the front strangle and make correction of errors. Then the class moves out, forming a double line all facing the instructor, and practices strangles for the rest of the 10-minute period.

[While practicing the two strangles, the instructor should keep reminding the students that they must get a firm lock on the enemy's throat so that he cannot break the strangle. Stress that in these strangles the important thing is to get the victim off balance. Mechanical strangles with a

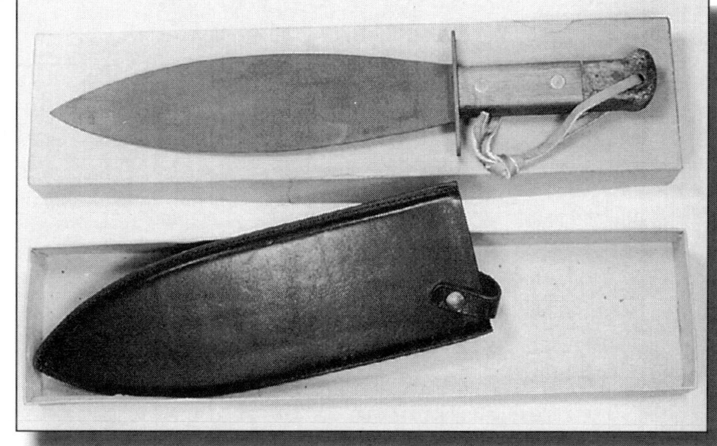

The bolo-style knife produced for the OSS based on the Fairbairn designed smatchet. An effective utility weapon, but apparently not used in many combat actions. (Photo courtesy of John W. Brunner.)

stick and rope are better. Give the proper approach. Here, the instructor will point out that the approach is very important and, naturally, that it must be noiseless. The attack should be launched from a leap of about 4 to 5 feet between the attacker and the sentry. The instructor points out that the first and most important objective is to strangle the man quickly and silently. In conjunction with the leap the fist will be driven into the opponent's kidney (with the right hand, if right-handed) with great force, causing the opponent to bend back and off balance. The instructor will emphasize this and explain why it is done.]

[When the students have gained proficiency in applying strangles, the group is divided into two lines. One line covers off the other at about 10 yards. One line acts as sentries and the other attackers. The attackers creep or crawl forward from the prone position to the point of rush and execute the strangle. This procedure can be followed in working out all sentry attacks.]

The instructor will reassemble the class and begin his instruction on sentry killing. He will tell the class that this hour we are going to take up sentry killing by mechanical means: with a stick, garrote, and knife. First, we will take up the stick strangle. Here it is pointed out that a stick of any kind can be a very lethal weapon, when properly used at the opportune time. Now, hold up the stick and grasp it about five inches from the end with the longest portion pointing toward the elbow and along the forearm. Then call out a student to be used in the demonstration. Get a man that is taller than the instructor. Next explain that you have to bring the victim down to your size. Accomplish this by a kick from the back at the junction of the knee or below causing the sentry to bend back and come off balance. At the same time bring the stick up with the right hand over and starting the longest end of the stick around the left side of the victim's neck and passing it under the chin around the throat. With the left hand reach up and secure the loose end of the stick. At this point your victim should still be off balance and pulling back on the student balancing him with the elbow and forearms; be careful not to exert too much pressure. Point out that the scissors formed by the arms pressing the head forward and pulling the stick back will cause strangulation quickly without any alarm whatsoever. Now the stick strangle is demonstrated without the numbers at normal speed. Answer any questions that might arise. The students move out and form a double line at extended intervals all facing the instructor. Proceed to do it by the numbers a few times and then change over and let the other line apply it to their partners. When the students have reached a certain amount of proficiency, they are called together and formed into a semicircle. It should be mentioned here that the hands gripped on the stick should be as close to the neck as possible. Show them how if this isn't done, it can easily be countered. Also, if the opponent is not off balance and reaches up with both hands on opposite ends of the stick, he can push up with his right hand and pull down with the left and pivot out of the strangle.

Now the instructor covers the garrote. The students are informed that this kind of strangulation can be accomplished by any type of cord or wire of strong tensile strength. The thinner the diameter, the more instant the effectiveness. Tie a loop for small wooden blocks at each end so that a secure grip can be taken. The length of the cord or wire should be no less than 18 inches. First note that you will watch the actions of the sentry, if possible. Again the approach is made from the rear, throwing him off balance by a kick against the inside of either knee with either foot. With a hand on each end of the cord, making sure that the cord is held taut throughout the operation, cross your arms at the rear of the neck and apply pressure both ways. Strangulation is quick and silent. The tightness of the cord or rope gives control of the weapon at all times, whereas if it were loose, it might land on the nose, chin or other areas but the throat. Now do the same method without numbers at normal speed. Again form the students into a double line facing the instructor. Men in the second row from the instructor secure the garrotes and will do it by the numbers a few times at the count of the instructor. Then change over and do the same process with the front or first row handling the garrotes for the remaining part of the 15 minutes. At this time the group is reassembled into a semicircle and more errors and fine points are pointed out, such as pulling the opponent back and off balance and holding the garrote taut.

Now take up sentry killing with a knife. Call a student forward to be used in the demonstration. In sentry killing all things regarding the approach and initial attack described for use with a strangle apply and, naturally, the approach is a noiseless one. The attack is launched from a distance of not less than five feet from the victim and is initiated as soon as the attacker has arrived at the spot. The upward thrust of the knife into the middle of the right or left kidney section is executed at the end of the leap to the attack. At the same time the free hand is clasped over the mouth and nose of the victim, pulling him backward off balance. The thrust into the kidney area has initially a great shocking effect and no outcry will occur if the free hand goes over the mouth and nose at the same time of the thrust. Pull the victim back upon the blade continually. After a few seconds, pull the blade from the back and, maintaining the same grip on the nose and mouth, lift the head up and back and slash the jugular vein.

[Men should be taught to hold the knife in the kidney until the subject ceases to struggle and then withdraw the blade and slash the throat. The tendency is to withdraw the knife and slash the throat much too soon.]

Now [the instructor] will go through the whole procedure without the numbers. At this time the students form the double line again facing the instructor. The second row from the instructor handles the knives and goes through the movements by the numbers a few times and then without the numbers. The instructor circulates around the students and makes corrections. Then change over and let the first row handle the knives. The process is done all over again until the end of the hour.

Then conclude that [in] this hour we have covered sentry killing by mechanical means—the stick, garrote, and the knife—and run through each one briefly to refresh their memory of all means.

MISCELLANEOUS BLOWS AND WEAPONS
[Document 9; 20 April 1945]

I. Instructors Needed
   1. One officer instructor per each two sections of enlisted men.
   2. One assistant instructor per each officer instructor.
   3. One officer instructor for officers.
II. Place to be Held
   1. Range or parade ground.
III. Equipment Needed
   1. Sticks, one stick per each two men.
IV. Preparation by Instructor
   1. Check with chief instructor.
   2. Check schedule book.
   3. Coordinate movements of class to area.

V. Miscellaneous Blows and Weapons (Pressure Points, Stick Techniques)

1. Pressure Points. Show the pressure points, describing how each might be used in the field. If time permits, demonstrate ear concussion blow.

[Ear Concussion Blow

Approach your opponent from the rear; concussion and rupture of the ear drums can be effected by cupping both hands and simultaneously striking them against your opponent's ears. A type of concussion which results causes the victim to become, according to the timeworn phrase, "slap happy," and makes him easy to do with as you will.]

2. Stick Technique. Show the use of the stick as a club and as a jabbing weapon.

3. Show other releases, such as blows to the testicles, finger breaking, stamping on insteps, bites, eye jabs; show kick with inside foot to shins, and, if time permits, demonstrate sitting neck break.

[If you are standing at the side of an opponent, clench your fist and strike him in the testicles with the hand on the side next to his body. This will cause him to bend forward for your follow-up, which will be an edge of the hand at the back of the neck or base of the skull.]

[Sitting Neck Break

If your opponent is sitting in a low-backed chair, approach him from the rear and as you pass by on the right side, at the point in which you are opposite, with the arm nearest the victim, reach across and under his chin with the hand coming to the back of the neck to break it instantaneously. It can be done almost without breaking your stride.]

4. Blackjacks. Show various forms of blackjacks, lead pipes, cables, and weighted socks and gloves used to subdue or render a man unconscious. Show areas of body that are vulnerable to this type of attack.

5. Brass Knuckles. Show the way they are used and concealed.

The subject of pressure points and stick technique will be discussed and demonstrated during the first part of this hour of instruction. The last 15 minutes will be devoted to a lecture on blackjacks.

There are numerous points on the body which will cause severe pain if certain nerve centers are pressed. However, they do not have any permanently damaging effect and can only be used to break holds. Other means mentioned before are better for this

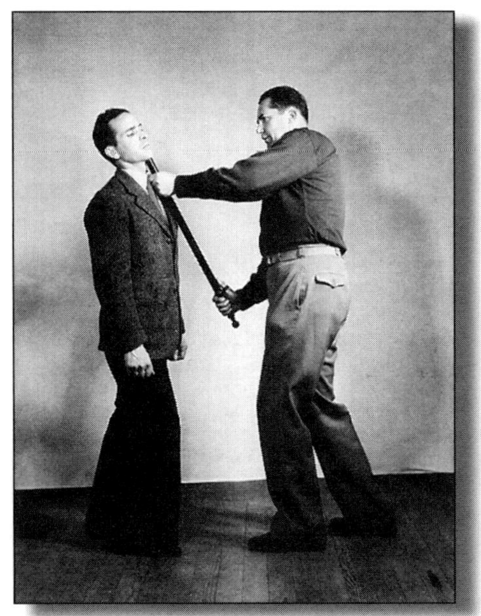

Applegate uses a club to demonstrate the advantages of armed versus unarmed defense.

purpose. However, due to a specific use, one nerve center is worth mentioning. If a man is lying on the ground, faking death or unconsciousness, and you desire to arouse him, lean over him and with your middle fingers press on each side of the head into the points on his skull where the jaw bone hinges. By pressing in and up toward the top of his head, you will cause such pain that no man who is pretending can stand it. He will come to his feet or give himself away instantly. (Demonstrate location of the points and tell each student to locate these points on himself. Inquire as to whether any student is unable to locate the pressure points.)

In conjunction with all the before-mentioned tactics, anything unusual or unexpected that can be done to confuse an opponent is desirable. If you can distract his attention, throw dirt in his eyes, or hit him with any object which comes readily to hand or create any mental diversion, you have placed yourself initially at a decided advantage. It is not a bad idea when anticipating rough and tumble tactics to have a small amount of sand in your pocket which may be thrown into a man's eyes, or to have a handkerchief folded in the breast pocket of your suit containing a little pepper or cayenne for use in your opponent's eyes. If in the midst of a fight you find yourself on top of a man trying to throttle him, you may hasten his end by beating his head up and down on the ground to stun his thinking process so that he may not try any of the numerous breaks to free himself from this position.

Knowledge of the correct use of a stick as a means of attack is very valuable. A man without other weapons is given confidence if he knows he can use it to take the offensive and down his opponent. Anything said here about a stick could be applied to a cane, umbrella, swagger stick or any other like object [club, night stick, baton]. A stout stick 18 inches long and 1 inch in diameter is about the minimum length and diameter with which the best results can be achieved.

The use of a stick in attack, combined with the element of surprise, is as follows: grasp the stick in the right hand near one end in a natural grip. Swing the other end up and grasp it about 6 inches from the other end in your left hand with the palm out. With a strong grip of both hands upon the stick and with your right hand held against your body so that this will be the pivot end of your stick, take the left hand and with force bring the left end of the stick across your opponent's middle section in a horizontal direction.

This blow, although not fatal and not always an incapacitating one, will give the effect of a solar plexus punch and will cause him to lurch forward with his chin out. Stop your horizontal blow at a halfway point across the opponent's stomach; from this position bring the end of the stick, which is in your left hand, up into the soft spot about 1 1/2 inches back from the point of the chin. At the time of the horizontal blow across the stomach, step forward or at least bend forward with the left knee to put more body power into the upward blow.

Naturally, if you miss the chin with the point of the stick in the upward jab, the other end of the stick will follow through and give a butt stroke effect as with the rifle. This particular technique ending with the jab in the underneath part of the chin will often kill, particularly if the point of the stick is sharp or if an instrument such as an umbrella or cane is used with a point or sharp ferrule on the end, which causes it to pierce up through the mouth cavity into the brain. (Demonstrate and then pair off students and practice technique by the numbers until proficient and then have them practice at will.)

Two other methods of using a stick in an attack are as follows: First method—grasp the stick in the same manner as described above with the exception that the stick is grabbed with the left and near the end <u>with the palm toward the body</u>. Facing your opponent and with a firm grip on the stick, sweep the stick upward, catching him underneath the chin to deliver a knockout blow. Follow through with this blow, bending your body backward at the knees as you deliver. Second method—with the stick held in the same manner, raise the stick in your arms to chest level and strike forward to the opponent's Adam's apple with great force. (Demonstrate both methods and have students practice.)

A blow struck at the top of the head with a club will not necessarily cause unconsciousness, but may break the stick. To disable a man using the stick as a club, it is best to use one of the following methods: strike a blow from a horizontal direction against the temple area of the skull. This will result in a fracture and a possible fatality. A blow delivered in a straight downward movement with great force at the point where the neck joins the shoulder will fracture the collar bone, causing the arm and the side of the body to be put out of commission. (Demonstrate and have students practice.)

If a man is to be put out of action and yet not seriously injured, a very effective way to do this if you are facing him or he is walking toward you (club in your right hand) is to push him on his right shoulder with your left hand, causing him to pivot, and at the same time deliver a hard blow on the back of the thigh across the large leg muscles, which will cause a leg cramp, incapacitating him for an indefinite length of time. This blow is used by police in mob actions, riots, etc. Police use the club as a jabbing instrument in most circumstances. They jab into the belly or solar plexus area to cause a crowd to give way, etc. [Mention was made of using a club in the same way as the fighting knife or smatchet. C.M.]

The balance of the hour will be devoted to a discussion of blackjacks and various types will be shown to you. The primary requisites of a good blackjack are as follows: it must have moderate weight in the striking end, the handle should consist of a spring affording flexibility, a retaining strap is desirable but not required, the handle is normally braided or woven leather which will improve grip, the entire assembly of a blackjack will usually be leather covered. If a commercial or [Counter Intelligence Corps] issue type of blackjack is not available, a "GI" sock partially filled with sand will do the trick very well. A portion of lead pipe may be inserted in a "GI" sock instead of sand. The various points on a man's body vulnerable to a blow from a blackjack are as follows: the temple, behind the ears, the back of the neck, the kidneys, the coccyx bone, and the back of the thighs. The use of the retaining strap around the wrist is debatable. However, it is well to remember that the strap is an excellent means of retaining your weapon if steps are taken by your opponent to disarm you.

A piece of zinc can be fitted to the knuckles and worn inside a pair of gloves and will provide you with a very effective weapon. Lacking the zinc knuckles, a piece of pipe may be inserted in the glove and used similarly as the blackjack. (Discuss the various types of blackjacks and show them to students.)

DISARMING
[Document 10; 20 April 1945]

I. Instructors Needed
1. One officer per each two sections of enlisted men.
2. One officer for instruction of officers.
3. One assistant for each officer instructor.

II. Place to be Held
1. Range or parade ground.

III. Equipment Needed
1. One M1 rifle with bayonet.
2. One Mauser rifle with bayonet.
3. One Arisaka rifle with bayonet.
4. One Lee-Enfield with bayonet.
5. One US Krag or Enfield per each two enlisted men.
6. One pistol per each two officers.

IV. Disarming

Split the officers' and enlisted men's work so that the officers practice most of the disarming with pistols, whereas the enlisted men use rifles. Spend plenty of time on the pivots so men can gain confidence. Give the men only one thing at a time, so you can maintain attention.

Stress the follow through to testicles and chin after muzzle is out of line with the body.

A few [minutes] at the end of the second period are used to give a familiarization with our own and enemy bayonet technique.

[Shall we take stock of our equipment? All right! Let's use "junked" revolvers, that is, weapons that have had firing pins removed and part of the trigger guard cut away. The reason for the latter move is to obliterate the danger of breaking a trigger finger in practice. Let's remember now to have a hammer that will fall on empty chambers to denote the passage of a shell. This, of course, will give us the necessary sound we desire. However, if automatic pistols are

Rifles and bayonets.

Disarming a firearm-wielding assailant was taken from an art to a science at MITC. It was considered a prelude to a follow-up attack on the enemy.

obtainable, that's fine. The point to remember here is never to attempt disarming practice with a barrel of less than 4 inches.]

The moment the subject of disarming is broached to a group of students, they immediately visualize two men furiously engaged in locked combat, each trying desperately to wrestle a bayoneted weapon away from one another. This type of fighting is basically taught in the infantry divisions and is entered into by all personnel. Naturally, the stronger wins. Then again, we find a practice entered into by a more highly trained group—the police. Here, we find our expert is intent on securing the weapon first and then having to properly subdue the enemy afterwards.

Perhaps a much simpler approach to this subject would be to remove the muzzle of the weapon from your body area and attack the man himself, striking a blow to a vulnerable spot and for a brief instant neglecting the weapon, but with the intention of destroying your man [or woman].

Let us first look at our pistol disarming. One can perhaps count on one hand the number of places in this country where pistol disarming is taught, so naturally the average layman knows little of this technique.

What does happen when a man walks up to you, shoves a weapon into your middle and says, "Hands up!"? Instinctively your hands go sky-high and you are startled. When ordered to "stick 'em up," you are expected to raise them to their full length. You may, of course, after a few minutes be allowed to lower your hands to where they are alongside the head, but under no circumstances will you be allowed to lower them any further. So remember, in practice, get closer to the normal situation by never allowing your student to start his disarming with his hands lower than his head. One thing is certain. If he had intended to kill you outright, he wouldn't have ordered you to "stick 'em up." So, he is saving you for some unknown reason. At this point, to start your disarming would be suicidal, since the mere "batting of an eyelash" might cause your opponent to fire the weapon. Men under this type of tension are highly nervous and as a result "trigger happy." Don't disarm immediately, but wait until you have taken your man's mind off his trigger finger. You can do this in any number of ways: ask questions, volunteer information or engage him in conversation. Once the man has relaxed his vigilance the element of surprise is in your favor.

Your first maneuver is (if the weapon is being carried in your opponent's right hand) sharply cut your left hand down, thereby straightening the arm at the elbow and striking the weapon clear of the body, keeping your feet in place and slightly bending your legs at the knees. To further safeguard your perilous position, it is wise to move your body out of range of the muzzle the instant the weapon is struck, thereby removing the danger of a luck shot coming close. Under no circumstances do you stop at this point, but move in, strike your man in the groin with a raised knee and, a split second later, deliver a hard chin jab blow to the point of his chin. You will note that after careful supervised practice you can escape the first shot and move in sufficiently fast enough to overpower your man. Naturally, those who are not too well coordinated must start their practice piecemeal. By this we mean let us first plan our practice to include only the simple body twist out of the line of fire and the bend at the knees. After successfully engaging in this for a time, concern yourselves with the follow through.

In your demonstration, it is wise to select a student for your opponent to avoid any idea of cooperation from the other's mind. Now have your men try it from the rear. Here, of course, your target is blind. But, here is an interesting point to think about. When a man comes up behind you and puts a gun in your back, you have a split second with your arms raised to glance behind you and size up the situation. Now you are familiar with two facts: the position of the barrel in the body area and whether your opponent is right- or left-handed.

Now that you have determined which hand holds the weapon, sweep down with your left hand (if your opponent is right-handed), straightening the arm at the elbow, and strike the weapon aside with the striking surface extending from elbow to fingers. Don't attempt to hit the weapon with your hand alone since your back is to your man. Once you have pivoted inside the gun arm, pin his arm to your left side by clamping your left arm over his gun arm. If properly executed, you will find yourself inside your opponent with his gun arm held securely and you striking [hard] blows both with your knee to the groin and your hand to his chin.

If at any time you are held up by one armed with a revolver, you can simply immobilize the weapon by grasping its cylinder. This prevents the trigger from being pulled. However, if the revolver is cocked on single action, the hammer will fall when the trigger is pulled in spite of the cylinder being held.

Let's consider another method of disarming. Here we have a slightly altered situation. Your opponent has held you up and behind him he has two or three cronies. Here you must secure the weapon to be used against the entire group. How do you do this? Let's see. In facing him, allow your left hand to drop on the gun barrel, engaging it with your fingers and at the same time moving your body out of range with the feet in place. Now with leverage in your favor, pull down on the barrel, thereby lifting the grip of the weapon against his hand, and at the same instant, slap the inside gun wrist sharply with your free right hand and at the same time force the weapon out of his hand. Once secured, step back two paces quickly, placing the grip of the weapon into the right hand, thereby covering the occupants of the room. This works especially good on long-barreled weapons.

Still another method of disarming from the front enables you to use the weapon as a club. When the gun is held in your opponent's right hand, sweep down with your right hand, grasping the barrel and simultaneously grasping his wrist with your left hand, then pivot to the right and wrench the weapon out of his hand. Maintain your hold on the wrist, as it will both encourage his losing the weapon and swing his body out of range.

We sometimes come across a student who says he is strictly a right-hander and can hardly use his left. We have something for him. Swiftly sweep your right hand against the inside of the gun wrist, securing it tightly, and at the same time pivot your body to the left, preventing a lucky shot from reaching its mark. Now, with the free left hand, bring it to bear under the weapon and force the barrel back against his body, in an attempt to break the trigger finger. Usually, the pain is so great that he is forced to turn loose of his weapon.

In our preceding examples of disarming we have dealt almost exclusively with the average inexperienced "GI" gunman, so let's take a look at the man who has had more advanced training in the proper methods of restraining an individual at the point of a gun. If your opponent has placed the flat of his left palm against you back, with the gun out of range at his hip, your pivot is to the outside, which brings you behind him. Securing your opponent's head in your right arm and lock his gun wrist with your left hand, taking him off balance.

The idea of pointing a finger into a man's back to simulate a gun barrel while holding a weapon with the other is fine, providing you bend the finger and place the knuckle there to fake a gun barrel.

A man who holds you up by keeping the weapon in his pocket is careless. If the weapon is in his right pocket, suddenly shove him sharply on the point of the shoulder of the gun hand, which in turn will cause his body to pivot so that the gun barrel points away from you. From here, move to his rear quickly or to his side, overpowering him, and trip him by placing your foot behind him.

Let us spend a few minutes on rifle disarming. Your chances of disarming a man with a rifle are good. Here you have a larger target to grasp. Besides, nearly all [World War II] military rifles are bolt-action weapons, and if you can escape the first shot, your chances of overpowering the gunman is good, since he will spend his time trying to re-cock the piece during the struggle.

Well, what can we do to escape the first shot? In our previous instruction we proved to you it was possible to move your body faster than a man can think to pull the trigger. Let's try it! With the muzzle of the rifle pointing directly at your middle, cut your left hand sharply away against the barrel, sweeping it to the left side, and sway your body to the right at the same time. Once the weapon's muzzle clears your body area, hold tight to the barrel (muzzle end) and cut away with your free right hand a sharp edge of the hand blow against the small vulnerable parts of the face or close your fist and smash at his eye region.

Or, let us use a different approach to this subject. Here we are facing our man again and we strike out with our right hand against the muzzle, grasping it and pivoting our body at the same time to the left. Reach over quickly and secure the weapon high up on the receiver with your left hand. Since your body has already pivoted toward the left, quickly lift out [the] right leg and strike the gunman sharply against his forward knee and simultaneously wrenching the rifle from his grasp and turn on him using the butt as a club against him.

From the rear (standing guard) straighten your arm at the elbow and lash against the muzzle and pivot inside...the weapon, striking with your knee at his groin and using the edge of your hand against his face or throat area.

Now take a walking guard. It is a normal procedure for any [guard] to want to prod a prisoner in his back while walking. It gives them an air of possessiveness, I believe. Instead of walking in a straight line, plant your left foot to the left of you at an angle of 90 degrees and pivot quickly. As his forward movement continues, your pivot will bring you to his rear where he can easily be overpowered. Make no mistake here. In taking a man from the rear, the weapon must be secured with the left hand while the right arm encircles his head, bringing him over backwards and off balance.

Unarmed defense against the rifle and bayonet. Too often we get reports back from the combat areas where we hear of infiltrating [enemy] suddenly rising from the edge of a cleared command post area and charging some high ranking officer, intending to bayonet him. Ordinarily one is so startled at the sight that "buck fever" envelops our man. When, if one could just have enough presence of mind to do so, one should drop into a crouch, raise the arms with fists clenched, fingers toward the face, and wait for the inevitable thrust. Depending on what section of the upper body area the thrust is aimed for, strike the flat side of the bayonet a sharp blow with the inside of the forearm, pivoting your body out of thrusting range. The forward movement of the attacker will carry him past your body and you will be close in to the inside of your opponent. Seize the upper handguard of his rifle with both hands, and since you have already pivoted your body to the left, strike him a sharp blow with the bottom side of the foot against the knee or groin and wrench the rifle out of his grasp or trip him by thrusting your foot in front of his feet, destroying his balance.

[A flamboyant technique, disarming was taught for those situations where agents were in extremis; about to be killed or worse. C.M.]

CHAPTER SEVEN

MITC: GO FOR THOSE FAMILY JEWELS

UNARMED OFFENSE
[Document 11; 20 March 1943]

1. Introduction

In addition to the invention of gunpowder and the compass, the Chinese are credited with being the first to develop a technique of unarmed combat. Chinese monks are reputed to have developed such a system to protect themselves against robber bands and nomad tribes. Over a period of centuries involving experiment, trial, error, and the loss of life, a system of defense and offense has been developed which has remained basically unchanged.

The Japanese, a short time after the 12th century, became interested in this type of combat, and, adopting Chinese ideas, began the development of their famous jujutsu technique. They gave it a mythological background, used it in training their young men, and have developed a form of religion based upon the application of its principles.

There were many variations of jujutsu taught up until around 1900 when a professor named [Jigoro] Kano, after study of all the various systems, established a school called Kodokan for the purpose of studying and applying this method of unarmed combat. The new system was called judo.

In the early 1920s a branch of the group using judo established itself in New York City. From that time on it has spread throughout the United States into the large cities, but

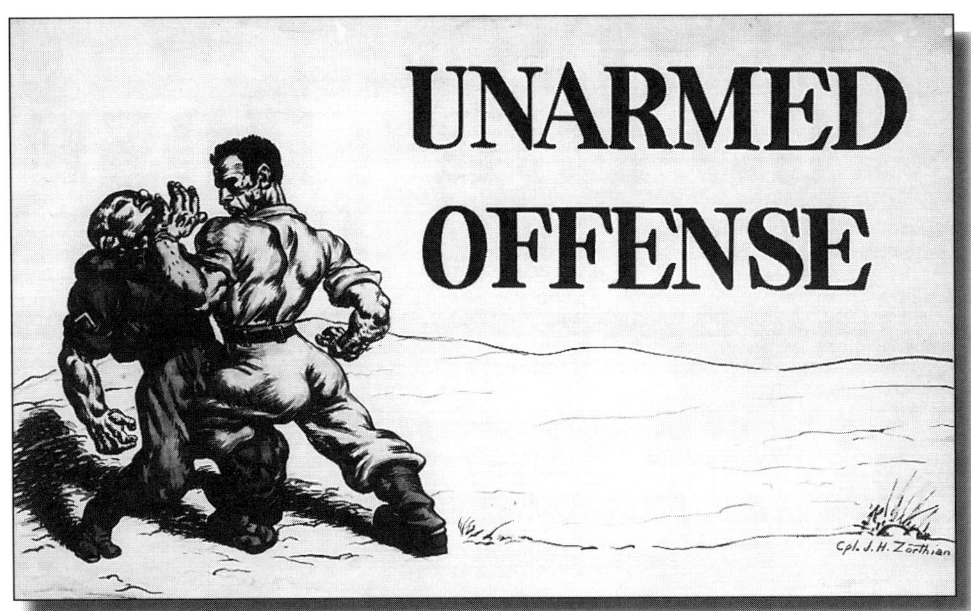

Unarmed offense. Training posters drafted by MITC artist Cpl. J.H. Zorthian to make the point that unarmed offense was a mental as well as physical state. These enhanced the image of American fighting men at the expense of the enemy.

has been practiced principally by Orientals. It has not gained much popularity among Americans, because, as in all things oriental, it involves a tedious amount of practice and a great deal of patience for the results obtained.

In recent years certain instructors set up schools which are directed toward attracting wealthy young men and are conducted in much the same manner as American rackets of the prohibition era. The pupil is told that he is being introduced into the mystic circles of the age-old secret methods of combat, brought down through centuries of Japanese culture, and, after paying a certain fee, he is instructed in some of its basic fundamentals. As time goes on and he becomes more interested, he is told that for an additional sum he can learn more holds and tricks. The results are very lucrative to the instructor.

Between this and the last war, numerous volumes have been produced by experts in judo, jujutsu, and unarmed combat techniques, all of which stressed defense as their sole purpose. The new army field manual on this type of combat, dated June 30, 1942, is titled *Unarmed Defense for the American Soldier* [FM21-150]. One should also be written titled *Unarmed Offense*. The instructions given over the past years to police and other law enforcement agencies have all stressed self-defense and restraint as the background of unarmed combat.

With the advent of World War II, public interest has been directed toward fighting and methods of combat. The desire of the American soldier and the American common man for knowledge of fighting techniques has greatly increased.

Present Day Status: Throughout the country today numerous articles are being written for publication in our magazines and newspapers, and illustrations of so-called "rough stuff" and underhanded tactics are being printed in the rotogravure sections of leading dailies. Throughout our armed forces various schools of instruction are being given by individuals who are qualified along orthodox lines and in many cases have had a great deal of police or other manhandling experience. The biggest trouble is that no one has outlined a uniform system of instruction or a constructive training course for unarmed fighting.

A confidence builder demonstrated to show how a smaller man can escape from the grasp of a larger opponent by knowledge of pressure points.

2. Fundamental Principles

All close combat techniques should be evaluated in the light of whether or not they will be useful to a man after he has lost his weapons.

Most of the combat which is being taught in our camps today is ineffective and not practical from the operational standpoint. This is because we lose sight of the fact that what a man is taught in this type of training must be simple, [the individual must be] able to execute it with great speed, and [it must be] practiced intensively until he instinctively reacts with the few blows, kicks, and releases that are necessary for his fighting education. Too many of the tricks taught to our men are useful only as a means of restraint. A good many instructors rely too greatly on the fact that an opponent will remain completely passive, without movement, enabling the student to apply a hold.

Anyone indulging in this type of combat cannot expect to get away without injury to his own person. Too many people

Zorthian, confidence builders.

have been given the impression that such methods of combat provide a magical means of subduing an opponent, without personal risk.

Simplicity: Any individual in combat in which his life is a stake very quickly reverts to the animal. There is a period of lucidity in his thinking for a few seconds (varying with the individual) and if he is well trained, he will automatically plan his offense. After a few seconds, and especially after he has been hit or jarred by his opponent, he reverts to the animal; the blood lust is aroused to the extent that from then on his combat is instinctive. Hence the need for a simple type of instruction with a great deal of emphasis on the few elementary methods which can be easily and instinctively used in combat.

Combat without weapons is characterized by two guiding principles: brevity and simplicity. Numerous texts have been written on this subject, but most people will find that they are rather complicated and involved and contain a good many useless things. We shall endeavor to reduce it to its lowest common denominator. We do not want to make a professional out of the average individual, but rather to teach a few simple tricks, which he can learn in a few minutes and use after practice.

Queensberry Rules Obsolete: Although the style of fighting which involves kicking a man when he is down, gouging out his eyes, and kicking him in the testicles does not appeal to the average American, we must forget the Marquis of Queensberry and his oft-violated rules of sportsman-like combat when dealing with enemies.

Ruthlessness is what we seek to achieve. It is best defined in two words: speed and brutality. In this type of fighting, it does not matter much what is done so long as it is done fast and as if life depended upon it, because it probably does. The two chief elements of success are surprise and speed. This applies as much to the individual as it does to the strategy of armies. We are aiming at attack alone and never at defense. Attack should be such that each blow will be a step toward or will itself end in a fatality. Each attack is also a defense.

This type of instruction teaches a man to fight and kill without the use of firearms, knives, or other lethal weapons. It is designed for use when those weapons have been lost (a situation which should be avoided at all costs) or when the use of firearms is undesirable for fear of raising an alarm.

Some of the blows described here are only a help to [that] end and should not be construed as a means to a certain and speedy demise of your enemy.

Developing Self-Confidence: The principles of unarmed combat are largely those of judo and various other styles of wrestling, boxing, Chinese boxing, self-defense and rough-and-tumble tactics. The importance of this type of combat lies not alone in the extreme offensive skill which the students can achieve, but also in the fact that any man, regardless of size or physique, once well trained in this technique, has a supreme self-confidence in himself and his fighting abilities which he could not achieve in any other way.

At some time or other all of us have been taught the rudiments of boxing under the Queensberry rules. These rules enumerate under the heading of fouls the principal targets which the boxer is not trained to defend.

At the present time we are at war. Our aim is to kill our opponent as quickly as possible. A prisoner is a handicap and a source of danger if we are without weapons. Forget the rules and use the so-called foul methods. These methods help to kill quickly. Hit the opponent in his weakest points. He will attack yours if he gets a chance. As a course of instruction of this type is designed to teach to kill, its practice and methods are dangerous without adequate instruction and supervision.

Any hold should be regarded as a means of getting a man into a position where it will be easier to kill him, and not as a means to keep him captive. The whole idea of releasing yourself from a hold or in applying one is to be able to kill. The disengaging move should form the beginning of an attack.

Balance, Mental and Physical: One of the basic fundamentals of unarmed body combat which must be firmly ingrained in the pupil from the very start is body balance. The man whose body is not perfectly balanced cannot utilize his strength, and his lack of balance can be used against him. The use of balance, as well as the use of your opponent's weight and strength when he is unbalanced, is one of the basic fundamentals of the famous jujutsu technique.

Body balance is best achieved by simple means. As it depends entirely upon the position of the feet, if the feet are kept the same distance apart as the width of the shoulders, balance is the result. Consequently, the feet must never be stretched wide apart or placed close together. It is advantageous to keep the knees slightly bent and arms hanging loosely at the sides.

In exciting circumstances, such as in combat, the condition of physical balance can only be retained by having mental balance. The first thing to do when on the offensive is to weaken the opponent's balance mentally and physically. Anything you can do to shake his mental processes may be a deciding factor. Yells, screams, grimaces, groans, etc., are all important. A push or pull applied to shoulders or other parts of the body weakens and breaks his body balance. Once off balance, his offensive powers or his strength cannot be utilized to any great extent.

In this manner a man who attacks first and throws his opponent off balance has a decided advantage regardless of difference in size. In this type of combat we hit, chop, thrust, poke or kick vital points on the opponent with the fist, elbow, the knee, the sides of the feet, and the heel and edge or palm of the hand.

Practice Essential: All types of hand-to-hand combat which demand set positions for the attacker and his opponent are useless when you find yourself projected into actual combat at an unexpected time. The only solution for those who have been students of this type of fighting is months and years of practice so that they can react instinctively to set positions of an opponent.

The other answer is the type of combat which we recommend. It does not depend upon any certain stance or position to achieve results. We must learn to hit not only here and now, but from all positions at the right time by acquiring skill in striking, poking, and kicking from constant practice on dummies.

To pick up a book on unarmed offense, no matter how well illustrated or clearly explained, is not sufficient to develop a fighter of this type. Actual demonstration by an instructor and supervised practice with other students is absolutely necessary.

Applegate demonstrates a balanced guard position for unarmed fighting that spreads the feet, lowers the point of gravity, and positions the hands and arms for immediate action.

In pairing off men for practice against one another, pair smaller men with much larger men. That is the only way in which a man of smaller stature can gain confidence in his fighting ability, because no matter how many times he sees these offensive techniques demonstrated by larger men, he will never have confidence in their use for himself. He will always subconsciously feel that "that's all right for those big guys, but how about me?"

<u>3. Unarmed Offense Techniques</u>

<u>Stay on Your Feet:</u> Don't believe that mastery of these elementary techniques will give you a magic formula which will bring you through all types of unarmed combat unscathed. Your main purpose always is to inflict as much damage as quickly as possible and at the same time receive as little as possible yourself. Two good rules in combat of this type are: 1) keep your opponent at arm's length and 2) never go to ground with your opponent, because this means getting too close to him, and being close, you will not have room to see what he is up to or to work yourself.

Although a cardinal rule in this type of combat is never to go to ground with or without your opponent, due to all conditions under which this fighting may take place, knowledge in the art of falling is very worthwhile. To be able to fall properly takes many hours of practice and such knowledge and

Zorthian, balance.

descriptions of methods can easily be procured through any book dealing with tumbling or allied subjects. As there is much difference between falling on mats and falling on uneven rocky ground, it is obvious that you should concentrate on remaining on your feet.

Let this one thing be said, however: once on the ground, never stop moving—start rolling and get to a position whereby you can easily regain your upright position. It is not necessary for you to follow your opponent to the ground once you have placed him there. Your feet are the weapon by which you can finish him. Remember that if you get to the ground and remain immobile, you are at your enemy's mercy and vulnerable to attack from his feet.

Blows Using the Hands: The best blows using the hands are delivered with the flat or heel of the hand or the edge of the hand. Blows struck with the fist (uppercut, haymaker, jab) are most effective when the user has had considerable experience in boxing and its allied sports. It will take up to 6 months to learn to deliver a positive knockout blow with either fist.

The heel-of-the-hand blow to the jaw (demonstrate) is the simplest and most effective of all blows of this type, and, when used in conjunction with a kick in the testicles, which

Zorthian, blows.

causes the opponent to bend forward, will often result in a neck fracture. The beauty of the last blow is its simplicity. In using the boxing technique, a man not used to using his fists may easily break a finger or cause a dislocation or cut himself on his opponent's bony facial structure.

Edge of the Hand: The most effective of all blows is the edge-of-the-hand blow. It is valuable because it can be utilized at vulnerable spots of the body which would not be so susceptible to blows from the fist or heel of the hand.

The edge-of-the-hand blow is delivered with the fingers extended, close together, thumb upright and wrist locked. The striking surface is the cushioned part of the hand between the base of the little finger and the edge of the palm where it joins the wrist. It is very important that the thumb be raised to an upright position because it prohibits the hand from remaining in a clenched position and the fingers automatically extend. The striking surface is well padded and its length, varying with the size of different hands, is roughly 3 inches. The thickness of the palm in most cases is about 1 inch.

Contrast the striking surface in square inches of this area with that of a clenched fist—where you have roughly 8 inches of striking surface. With the edge of the hand you have only 2

Blows delivered by open hand and knees were preferred to harder-to-learn fist strikes.

or 3 square inches. Therefore, a blow delivered in this manner gives a sharp-edged effect, causing a break, fracture, or concussion because of the force expended on a relatively small area.

The blow should be delivered with the elbow bent, utilizing body force behind it with a chopping motion (demonstrate). The last is important because it tends to localize the force of the blow in a small area. If you deliver the blow and do not quickly draw back your hand from the part of your opponent's body attacked, a great deal of the striking power is expended over a larger area and thus becomes less effective.

This blow can be delivered with either hand in a downward direction or horizontally with the palm down as in a backhand sabre [tennis] stroke. The best position in which to use the horizontal edge of the hand blow is with the right foot forward, and with the favorite hand (usually the right). In this position your body weight can be utilized more fully. The reverse foot position applies for the left hand. With a somewhat lessened effect, the blow can be delivered with either hand and from any free position where the arm can be swung.

<u>The Chin Jab:</u> This extremely effective blow is so named because its only use is on the chin area. It must be delivered up and under the chin with the heel of the palm, fingers extended to give the palm rigidity. The more directly underneath [the chin] it is, the more power it will pack. It is executed with a bent elbow, and a great deal of body force can be utilized at the time of impact (demonstrate).

The further forward the chin is extended at the time of the blow, the more devastating the result will be. If a knee thrust to the testicles or groin is used in connection with a chin jab, the body will be automatically bent forward, leaving a perfect setup for this particular blow. It results in unconsciousness and possible neck fracture if delivered with sufficient force.

The arm or hand does not have to be drawn back [for] execution of the blow, but can be hanging at the side, fingers hooked in belt, hand resting on lapel, or in any nonchalant position. An average man can cause a knockout with only 6 inches of traveling distance from the start of the blow to the point of impact. The element of surprise is most useful in close quarters where time, space or circumstances do not allow the hand and arm to be withdrawn for a long haymaker.

A neck fracture can be caused by gripping your opponent's belt with the left hand and jerking him forward at the moment of impact with your right. It is also desirable to use the fingers of the striking hand on the eyes following the blow.

The elbow also fits into the inventory of blows available for use to close-in fighters.

Chin Jab and Trip: If you wish to down an opponent while passing him on the street, utilizing your advantage of surprise, this is a very simple and effective means which can be executed without any suspicious warning movements. As you pass your opponent on the left side at the time in which you are directly opposite him, place your right leg in the rear of his right and execute a chin jab from a position of hands at side. He will go down and out. The leg in the rear has the effect of causing the body to go up and come down with more force.

Ear Concussion Blow: Approaching your opponent from the rear, concussion and a rupture of the eardrums can be caused by cupping both hands and simultaneously striking them against your opponent's ears. A type of concussion which results causes the victim to become "slap happy" and makes him an easy subject to do with as you will.

Use of Fingers, Elbows, etc.: An eye gouge, a bite or lip tear is always good at close quarters as a means of disengaging your opponent's hold or weakening him before the final finishing off process. The eye gouge is best accomplished by placing a thumb on the inside of the eye socket next to the nose and flicking the eyeball out toward the edge of the cheek. However, in itself it is not sufficient to permanently put a man out of action once his fighting instincts are aroused. The same is applicable to the lip tear—hook your thumb in the corner of the mouth and tear toward the hinge of the jaw.

Zorthian, kicks.

Using the elbow against the midsection or other tender parts of the anatomy is always good and very effective, as well as such other stratagems as stamping on the instep, kicking the shins, pulling hair, breaking fingers. One or a combination of these things is sufficient to effect release from nearly any encumbering hold, such as a grappling hold from the rear or the front, or a standing front choke hold. Any choke hold can be broken if you can grab one of the fingers and break it.

Kicking: A kick delivered toe foremost and aimed at a narrow target is not accurate enough, particularly when the slightest move on an opponent's part will cause you to miss and leave you off balance and wide open to his retaliation. [Emphasized over hand blows in later years. C.M.]

Unless you have unusually good footwork and balance it is not advisable to try to kick a standing man at any point above knee height unless his hands are otherwise engaged. The savate or French method of fighting with feet is a very difficult art to master and, if faultily employed, can result in disaster to the attacker.

Kick with either the inside or the outside of the foot. This blow delivered with the aid of heavy footwear gives a

Applegate demonstrates the advantages kicking has over the reach provided by using the arms for blows.

striking surface of the length of the foot from heel to toe. When properly delivered, it does not leave you unbalanced in case of a miss or a near miss (demonstrate).

The kick should be delivered from the front directly to or a few inches below the knee cap. Allow the foot to scrape down, putting the weight at the finish across the ankle joint. This has the effect of bruising the tender shin bones and crushing the small bones of the foot, which are very tender and unprotected.

If this blow is delivered properly, the knee will go out, or if the knee is in a slightly flexed position, not getting the full benefit of a blow against its hinge, the foot will be crushed and rendered completely useless by the follow-up down the shin. The effect is to cause your antagonist to topple to the ground, leaving him open for an easy kick to the rib area or temple after he is down.

The same kicking blow delivered against the knee from the side will have the same effect as the one from the front. At any time a kick on the shins will cause the strongest man to lurch forward and stick out his chin, which leaves him open for the chin jab or uppercut.

Kicks as Coup de Grace: After your opponent has been downed, the coup de grace can be given with the foot in the form of a kick. This can be done either with the toe of the foot or by driving the heel into the midsection, throat area, or temples with great force. In either case, it is best to be wearing heavy boots or other heavy footwear.

When using the heel to finish off your opponent, it is best to use one foot only, driving it into the rib section or other vulnerable spot. Thus you may more easily maintain your balance than if you jump on your opponent with both heels. This latter method of using both legs is particularly dangerous if the ground is uneven or the fallen man rolls,

Zorthian, strangles.

because you are apt to lose your balance and go to the ground with him.

Japanese Strangle: A lot has been said concerning various types of strangle holds, principally for use in wiping out a sentry by an attack from the rear. It is obvious that in this case a knife is desirable.

If you must accomplish the task with your bare hands, the following basic fundamentals should be remembered: The approach must be a noiseless one. Your attack should be launched from a leap over the remaining 4 or 5 feet between yourself and the sentry. This leap is important because a great many people, especially when they are on the alert, have a super developed animal instinct which gives them warning of a hostile presence, although they do not see or hear anything.

Your first and most important objective, of course, is to strangle the man quickly and silently. In conjunction with your leap, your fist should be driven into the man's right kidney section with such a force that he will be caused to bend backwards and thus come off balance. At the same time, your left forearm should be swung around his neck in such a manner as to strike him across the Adam's apple. These two blows are enough initially to stun him for the few vital seconds, which are necessary to apply a quick, scientific strangle hold (demonstrate).

From this position with your left arm across his neck, place your right hand on the back of his head and hook your left hand inside the bend in the elbow of your right arm. With your hand in this position, you are able to exert enormous leverage by pushing forward with your right hand and pulling back with your left at the same time. In a matter of seconds you have strangled him completely or broken his neck.

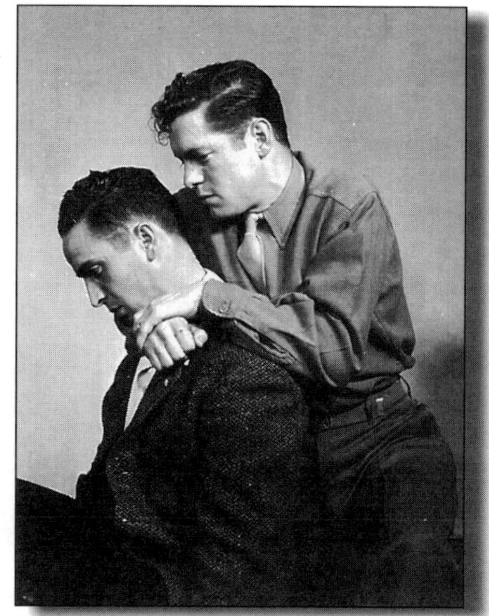

Strangulation was a means of exploiting surprise to overcome an enemy, especially if armed.

One of the most important things about this particular hold is that you must be continually pulling your victim backward so that he is off balance at all times. This is even more important in case you are shorter than your victim. In this case the use of the knee instead of the fist is the best for the first blow.

Front Strangle: A strangle hold may also be applied from the front in the following manner. It is easier when a man's head happens to be lowered as it would be if he were attempting to make a grab for your legs and waist. If standing, swing your right arm forward and around bringing the palm of the hand against the back of his neck. By giving your body weight to the swing, you will cause him to bring his head forward and downward to a position where your left arm can be brought up and under his throat and locked around his neck, with your right hand taking a grip on your left as reinforcement.

When you have him in this position, all you need to do to cause strangulation or a neck break is to push your hips forward and your shoulders well back, lifting upward as you do so.

Sitting Neck Break: If your opponent is sitting in a low-backed chair, approach him from the rear and as you pass by on the right or left side, at the point in which you are opposite him. With the arm nearest to the victim reach across and under his chin with the hand coming around to the back of his neck. From this position, a contraction of the arm muscles plus an upward and backward jerk will cause his neck to break

Zorthian, come-alongs.

instantaneously. It can be done almost without breaking your stride.

If you are standing at the side of your opponent, clench your fist and strike him in the testicles with the hand on the side next to his body. This will cause him to bend forward for your follow-up, which will be an edge-of-the-hand blow at the back of his neck or the base of the skull.

4. Miscellaneous Holds, Throws, and Ties

Come-Alongs: The subject of "come-alongs," or a means of bringing in a captive, is a long and varied one, but no such hold yet developed that is applied by bare hands can be maintained successfully over long periods of time without being weakened to such an extent that it may be broken by the captive. It is true that in some escapes from these come-alongs the victim will hurt himself, but at the same time, if he is desperate enough, that will not deter him from an attempt to escape. The only way to keep a prisoner being taken in by come-along methods over any distance with no danger of his escaping is to keep him in a perpetual state of semi-consciousness by edge-of-the-hand blows to the neck, short jabs on the chin, or by any means which will keep his mental processes foggy. It is well to take his free hand and stick it down inside his belt to prevent possible escape attempts.

The most effective of all come-alongs, particularly when you are forced to walk a man a long distance and keep him under control, is the following. Since the come-along is not an attack, you have already subdued him to a point where he is submissive.

Facing your opponent with your right hand outstretched, palm up, grasp his left hand on the back, with your hand holding across the back of his fingers. With your left hand, reach over on the outside of the prisoner's elbow and pull it toward your right foot until he is directly opposite you (demonstrate).

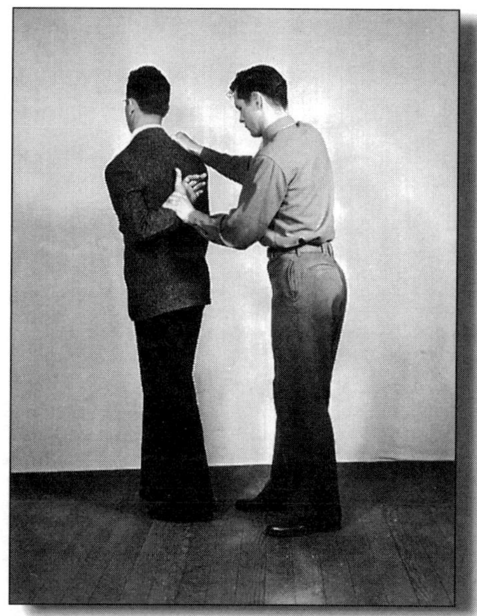

Come-alongs were used to move prisoners short distances until they could be secured with ties or otherwise neutralized.

You will find that the victim's left elbow will be next to your body with your right elbow between his arm and his body. You have not changed your grip from its initial application during this process. By keeping his elbow close to your body and locked in place by your right arm and raising the forearm to a vertical position, you have a very effective come-along which is maintained by twisting his hand and wrist toward you at any sign of rebellion. By applying a few pounds of pressure on the wrist, you can raise the victim on his toes, and it is by this means that you will know that he is completely under your control.

This come-along has the advantage of allowing you, in most cases, to be able to maintain sufficient pressure with one hand to enable you to walk along with a weapon or some other implement in your left. This application can be reversed for the purpose of leaving your right hand free if desirable. Ideally the initial grasp on the victim's hand should be done by hooking, as illustrated. Due to poor light, gloves, weather, etc., this cannot always be done easily. You can change to the thumb position once the hold is applied if you so desire.

Another come-along which has a great deal of merit is the arm lock. Properly applied, this lock makes a hold sufficiently strong for escorting a prisoner a short distance. It gives control of your opponent completely if pressure is maintained on the forearm, and it is very useful as a [compliance] hold or

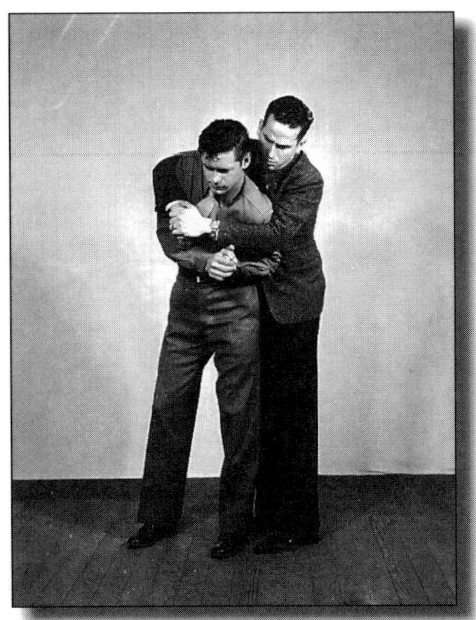

If caught, kicks-blows-throws were available to release and set up a more lethal attack.

in taking a man to the ground before tying him. The proper speedy application of the [arm] lock can constitute an attack. However, it is easier applied as a mastering hold after your victim has been subdued by other means.

Its application is as follows: Facing your opponent, reach out with your left hand, palm down, and grab the opponent about the right wrist. Shove his arm to the side and rear of the body. In conjunction with this move take the flat of your right hand and strike the left arm on the inside of the elbow joint. The hand should be immediately withdrawn after the slap has been given, causing the elbow to bend. From this position disengage your left hand, which has been about his wrist, and shove it under and up between the opponent's forearm and his back. Place your left hand on or just below the point of the shoulder on his arm (demonstrate).

By bending forward with his right arm locked in this position you have him completely under control. Your right hand can then be used to exert extra pressure on his pinioned arm by pulling it out from his back. This will force him to do as you will because of the pain or possibilities of a broken elbow. This come-along can be maintained over a long distance but has a disadvantage in the fact that your own body must be bent forward alongside and slightly over your opponent's body to keep him under control.

<u>Throws:</u> There are any number of throws which could be described, but one of the most effective and simplest is the good old "flying mare." It can be applied swiftly by grasping your opponent's right wrist with both hands, stepping in with your right foot, turning your back to his body, and bringing his arm over your right shoulder with the hinge joint of his elbow up (demonstrate). In this position, you will have a firm grasp of his arm on which pressure will be exerted against the hinge resting on your shoulder so that any sudden downward movement of your body combined with a quick back thrust of your hips will send him sailing through the air. If he doesn't go, his arm will break from the leverage exerted, and he can be finished off in some other way.

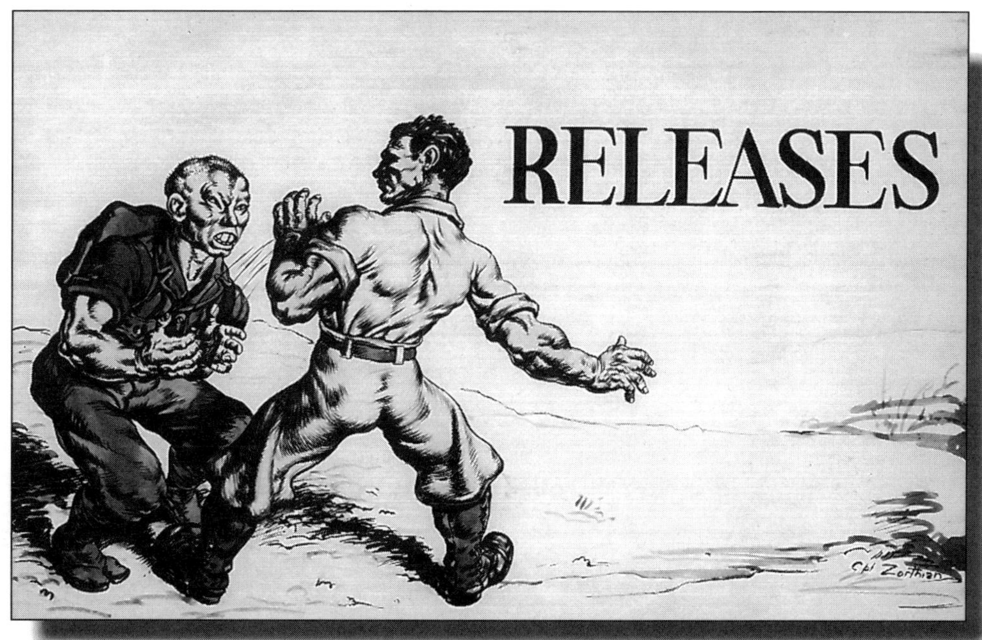

Zorthian, releases.

If you are working on a hard surface, in place of letting go of your opponent, flip him over on your shoulder, maintain your hold upon his arm after the throw, and bring him down on his head and shoulders at your feet with a resultant concussion or neck fracture when he strikes the ground.

Wrist Throw: The wrist throw should be mentioned here because it has several practical applications. Its most practical use would be in a situation where a man has reached out and grabbed your shirt or coat lapel with his right hand. With your left hand reach over and to the inside of the grasping hand and place your left thumb in the back of his hand between the small knuckle bones of his first and middle fingers. Your fingers will pass underneath the palm of his hand. With your hand in this position, twist his hand sharply back toward him and to his right and force it toward a point on the ground three or four feet to his right (demonstrate).

He will immediately be forced to go to the ground and from there you can release your hold upon his hand, pulling his arm straight about his head as he goes down, and kick him in the temple with your foot. In many cases, particularly when there is a great difference in size of opponents, it is advisable after making the initial hold with your left hand to use your right to give additional pressure and leverage to complete the throw. The same instructions can be applied by doing just the opposite in the case of a left-handed procedure.

Wrist Release: Mention should also be made of the principle of a simple wrist release as it is an invaluable aid when someone has grasped you by one or both arms. The first thing an opponent does is to grab either your wrist or forearm. This is obvious because he wants to protect himself as well as immobilize your own offense.

When a man grasps you by the wrist, he will have four fingers on one side of your arm and the thumb on the other side. Regardless of how strong a man is, the thumb, which is the weak side of his grip, will not be stronger than your entire arm. By a twist of your wrist outward against the thumb, you can break his hold with a sudden effort.

The entire movement must be rapid. If you will always twist your wrist against your opponent's thumb, regardless of whether it is his left or right hand, you will be able to break his grip. If he grabs your wrist with both hands, by jerking upward toward the thumbs the same release can be effected with a little more effort.

Pushing Counter: Every man at some time or other has been in a position where a belligerent opponent or drunk has attempted to antagonize him by placing a hand on his chest and shoving him backwards. The counter is simple and effective. As your opponent's hand is placed on your chest, take your own two hands, laying one flat on top of the other, raise them above your opponent's pushing hand, and come down sharply with the edge of the hands at the angular bend where his wrist joins his hand. As you do this, bend forward (demonstrate).

It is important that you bend forward in applying this at the time of the blow on the wrist angle. By doing so, you force him to the ground and also pin his hand against your chest in such a manner that he cannot pull away.

Your opponent will go down for one simple reason. When he is pushing you, his wrist is already at a right angle. Any additional bend will cause a break. When you strike his wrist with the edge of your hands he can do nothing but go to ground to protect himself from a broken wrist.

As he goes down, you can use the knee in his testicles or chin or do whatever you will, depending upon how you desire to dispose of him.

Ties: The following method is a means of tying a prisoner

securely. It is initiated from the arm lock with your opponent's face downward on the ground and with the forearm bent up behind his back in a painful position. A little additional pressure on his arm when in this position will force him to readily place his other hand behind his back at your request.

With your rope, tie the two wrists tightly together. Take one end of the cord, run it around his neck, and tie it against the pinioned wrists. Have enough pressure on the cord around the neck to force his hands high up toward his shoulder blades.

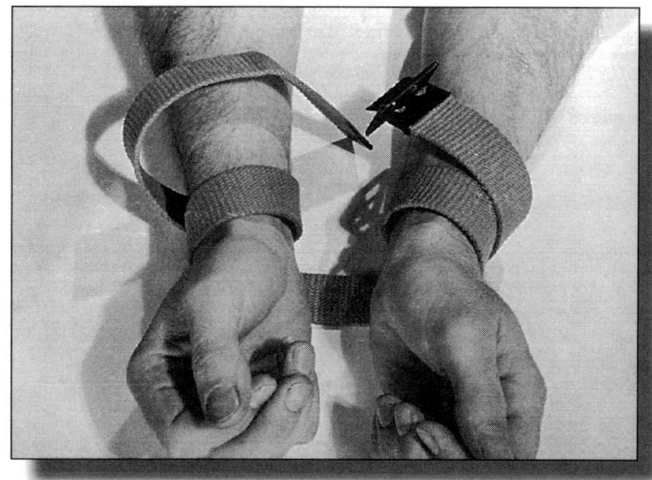

Using a "GI" web belt to secure a prisoner, one of a number of ways to reduce the threat of escape or attack.

Cross his ankles and take the other end of the cord after doubling his legs behind him and tie them with it so that they remain in that position (demonstrate).

Any struggle to free himself will result in strangulation. Correctly applied, there is no escape from this tie.

There are various knots advocated in making such ties, but any standard tying knot may be used. The essential thing is the fact that the victim will be unable to make any effort to release himself, regardless of the type of knot used.

Another simple tie can be effected by using a man's shoe laces for cord. Place him with his back to a tree or post, preferably of 10 or more inches in diameter. Have his arms placed around and behind the pole, tie his wrists, two thumbs, and two little fingers together with the shoe laces. In these ties, practice is most important, because a faulty procedure in tying is glaringly apparent.

5. Vulnerable Spots of the Body

A man's body is made up of many parts; some are hard, some bend, some do not, in some places nerves are near the surface. The following are a few of the weak points of a man's anatomy which are most vulnerable to attack.

The Crotch: The testicles are the most vulnerable and sensitive part of a man's body. Any foot or hand blow delivered in the crotch will enable the weakest man to knock

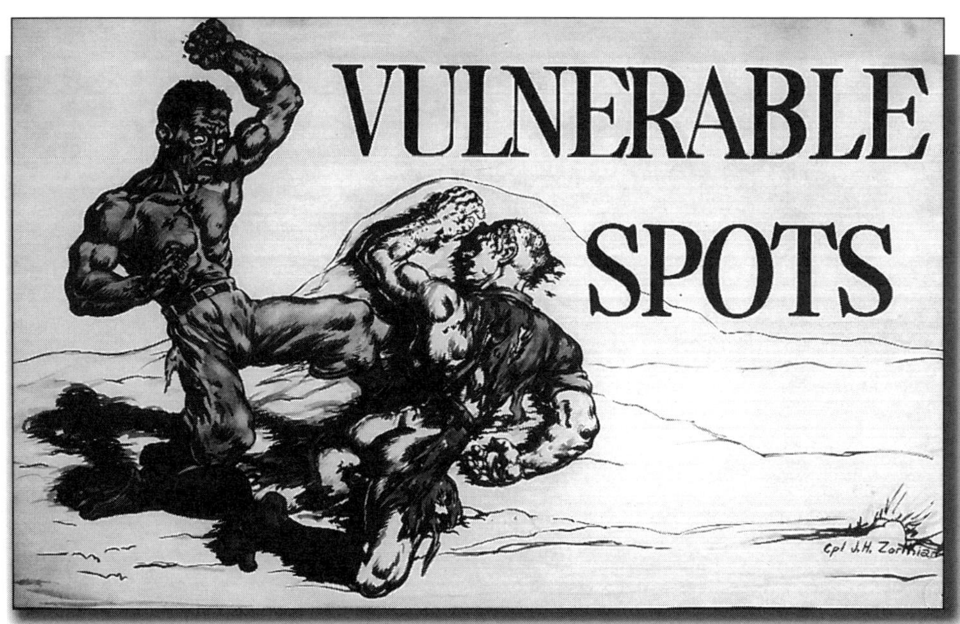

Zorthian, vulnerable spots.

the strongest man senseless or to disable him to a point where he is easily finished off by some other means. The strongest holds can be broken at any time by grasping an opponent's testicles and pulling and twisting them. Never forget this Achilles' heel of a man's anatomy.

The Chin: The time honored punch on the jaw is hard to beat. Delivered by a skilled boxer it puts a man down for the count. The force of the blow on the point of the chin causes a form of concussion and we have the so-called "knock-out." This same result can be obtained by a blow from the heel of the hand, which packs a terrific wallop. Further, it is much easier for those inexperienced at boxing, it has more of an element of surprise, and it can be used from a shorter starting distance.

The Windpipe: A blow with the edge of the hand across the windpipe causes temporary if not permanent blackout. The blow has the same effect as crushing a piece of copper tubing with a sharp edged instrument. The effect of such a blow on the windpipe can be easily demonstrated by having someone place his thumb in the small hollow at the base of the throat, [and] pressing gently.

Blows on the sides of the throat and on the large cords at

the back of the neck will cause dislocation, concussion and very often a break. Very few physiques will stand edge-of-the-hand blows on these spots.

<u>The Bridge of the Nose:</u> An edge-of-the-hand blow delivered at the point where the nose joins the bony structure of the brow causes a crushing of the most fragile part of the bony structure and brings unconsciousness and possible death from cerebral hemorrhage.

<u>The Nose:</u> A blow with the edge of the hand directed underneath the nose in an upward direction toward the forehead causes a crushing of the frontal bones, unconsciousness and cerebral hemorrhage.

<u>The Kidneys and Small of the Back:</u> A glance at a physiology book will show that the main muscle cords and nerves of the body branch out from the base of the spine at a point very near the surface. In this area, a sharp sudden blow has a great stunning effect. The entire section across the back 6 inches over the base of the spine, including the right and left kidney, is sensitive to this form of attack.

<u>The Navel and Solar Plexus:</u> One inch below the navel is another vulnerable spot, which, if hit by a finger jab or other sharp blow, will cause unconsciousness. However, it is not easy to find this spot. It must not be confused with a blow to the solar plexus, which is delivered above the navel in and up under the rib structure.

<u>The Knees:</u> The knees, because of their hinge type of structure, are particularly susceptible to hard blows, especially those struck by the feet. If we kick a man from behind on the back of his knees, he will fold up off balance and may be polished off that much faster. Blows or kicks delivered directly at the knee cap from the front or directly from the right or left side of the knee when the leg is straight will cause a break. Even light blows delivered in this manner cause dislocations of tendons and nerves, such as are common among football players, known as "football knee."

<u>The Arms:</u> The arm has three joints: the shoulder, the elbow, and the wrist. Nature made them bend one way only. If you force them in the opposite direction, they will either break or cause your man to go down.

There are other vulnerable spots, but a knowledge of those mentioned above is sufficient for unarmed combat purposes.

[Neck Pressure Points: There are numerous pressure points on the body which will cause severe pain if certain nerve centers are pressed. They do not have any permanent damaging effects and can only be used to break holds. Other means mentioned before are better for this purpose.

However, due to a specific use, one nerve center is worth mentioning. If a man is lying on the ground, feigning death or unconsciousness, and you desire to arouse him, lean over him and with your middle fingers press on each side of the head into the points of his skull where the jaw bone is hinged. By pressing in and up toward the top of his head, you will cause such pain that no man pretending can stand it. He will come to his feet or give himself away instantly (demonstrate).]

6. Conclusion

In conclusion, it is well now to emphasize again the fact that in applying the numerous holds, come-alongs, and throws that are being taught, you cannot always assume that you are against an untrained adversary, particularly so in the case of the Jap. If this is not the case you will lay yourself open to attack while attempting their execution.

If you are in a position to apply any of these encumbering holds, you are also in a position for a killing attack initiated by hand or foot blows.

In conjunction with all of the before mentioned tactics, anything unusual or unexpected that can be done to confuse an opponent is desirable. If you can distract his attention, throw dirt in his eyes, or hit him with any object which comes readily to hand or create any other mental diversion, you have initially placed yourself at a decided advantage.

It is not a bad idea when anticipating rough-and-tumble tactics to have a small amount of sand in your pocket which may be thrown into a man's eyes, or to have a handkerchief folded in the breast pocket of your suit containing a little pepper or cayenne for use in the opponent's eyes.

Dummies are absolutely necessary in training for this type of combat. They should be of standard size, complete with arms, head, and legs. The vital spots should be marked on the dummies and the student should be made to practice daily with no restraint all hand and foot blows he has learned.

It is easy to see that if in practice of this type of combat you use a sparring partner, great care must be taken in application of this instruction. Submission signals should be arranged to avoid injuring each other.

A man who masters the blows emphasized and practices them enough to be able to use them as readily as he uses his fists need fear no one. Even if he never has a chance to use the technique in actual fighting, it will still be worthwhile because of the supreme self-confidence he will develop.

CHAPTER EIGHT

SINCE 1945

At war's end, Lieutenant Colonel Applegate briefs Secretary of War Henry L. Stimson on what was accomplished at the MITC, which contributed to debate on postwar national intelligence organizations.

Lieutenant Colonel Applegate at the Camp Ritchie "Bavarian" village training area after the war. While the organizations that worked on close combat were dismantled, he took his knowledge and experience to other audiences.

The final Allied victory in Europe and Asia marked the end of many enterprises involved in this story. On a general historical note, the significance of the British and American work performed behind enemy lines should be acknowledged. The OSS and SOE partnership was by no means universal and both faced competition from other so-called allied institutions (in the United States this was with the FBI in North and South America and with the existing MID of the War Department and the Office of Naval Intelligence of the Navy Department in Europe, Africa, and the Pacific).[1] But without the training, materiel, communications, and leadership by the SOE and OSS, local resistance would have had little military value.[2]

The OSS was dismantled in October 1945 and the SOE in January 1946, which was followed later that year by the War Department's MITC, which split its functions between Fort Riley, Kansas, and Fort Holabird, Maryland—without a Section Eight. Promoted to lieutenant colonel, Rex Applegate medically retired from the U.S. Army. For efforts with the OSS, Fairbairn also made lieutenant colonel and was awarded the Legion of Merit when he returned to England. The unheralded Major Sykes had passed away at war's end, while Fairbairn went on to train police in Singapore and Cyprus.

Applegate recalled that had it not been for World War II, Fairbairn and Sykes "might well have lived out their retirements in relative obscurity in the Far East. Then the modern fighting man would never have benefited from their experiences and skill in hand-to-hand combat. Both men left a legacy of close-combat techniques and training methods that are in use to this day."[3]

In varying quality and quantities, the U.S. armed forces included portions of their gutter fighting in most training curriculums, using available instructors and published material. Author William L. Cassidy concluded that by 1945, the U.S. armed forces had been exposed in some degree to close-quarter combat, and if Fairbairn and Sykes are "called the great innovators of combat shooting, then Rex Applegate must certainly be its finest instructor."[4] The three "gutter fighters'" greatest impact was with special operations personnel, who ranged from James Jesus Angleton to Henry Kissinger. (At one time, an estimated 13,200 people served with the SOE; 12,718 with the OSS; and another 500 with the AIB. By various estimates, 14,696 servicemen passed through MITC.)

The joint service nature of the SOE and OSS allowed for cross-training, but their clandestine character also hindered enduring development as methods disappeared into the black hole of need-to-know postwar security. At the MITC, this SOE/OSS approach was perfected by Applegate's Section Eight team for mass-produced operatives and agents regardless of background, gender, ability, and ultimate assignment. Of the various versions of the Fairbairn and Sykes techniques transferred to the United States during the war, Applegate's stands out as the best example of organization and instruction for general-purpose wartime needs.

Applegate felt that the average American lacks the time, patience, and usually the interest to become a genuine expert, but "can be quickly turned into a dangerous, offensive fighter by concentrating on a few basic principles of combat and by advocating principally the use of blows executed by the hands, feet and other parts of the body"—and can be made ready to use firearms, bayonet, knife, and any improvised weapons before reverting to unarmed combat. Fairbairn added not to consider yourself an expert "until you can carry out every movement instinctively and automatically."[5] A contemporary martial artist observed that World War II SOE and OSS "close quarter battle training was very short, very rudimentary, but very intense, and focused on scenario-based training. What is amazing is that these old-timers, who are in their 70s and 80s, are still able on demand, to recreate and utilize those combative shooting skills they learned years ago."[6]

The basic outline of a quick, dirty, and direct program remains today as a close-combat system that includes armed, unarmed, and firearms for direct contact with the enemy. More complex and precise methods exist, but these assume that time, talent, and facilities will be available to train expert athletes. Wartime and peacetime constraints belie this and what was presented during the war were a set of basic skills taught in reduced hours of instruction. Underlying principles of this approach include surprise, speed, range, intensity, and relative advantage. These methods both physiologically and psychologically fit the high stress that exists by subscribing to the keep-it-simple-stupid principle for use under tension and for brutal effectiveness—kill or be killed.[7] Applegate's applications of these methods can be found in the various editions of his works published since, and this book provides original wartime material without adornment.

Asked how much impact the close-combat training conducted by the OSS and MITC had in postwar law enforcement, clandestine operations, or the armed forces, Applegate acknowledged: "It all seemed to have disappeared by the 1950s. Most agencies subscribed to a 'not invented here' attitude. The competitive marksman and martial artists were not interested in anything that took away from their chosen art."[8]

Practical combat shooting techniques with handguns and shoulder weapons did not transition well to peacetime, although they are documented in journals and training material.[9] For example, the National Rifle Association (NRA) commented in an article on point shooting that "the methods advanced do not necessarily reflect NRA opinion."[10] Prewar competition and law enforcement institutions were still in place with concerns for scores and safety and an American prejudice for revolvers. Shooting and unarmed forms of close combat once again divided into dissimilar fields as commercial and collegiate combative sports moved back into the arena. The wartime aspects of armed and unarmed close combat seemed only useful to combat soldiers or gangsters.

RIOT CONTROL, FIGHTING KNIVES, AND COMBAT SHOOTING

Applegate moved to the private sector in 1945 but continued to be interested in the security field. In Mexico City he managed a Nash automobile assembly plant until 1947, but turned to the import business with the Sanborn family as president of Cia. Importandora Mexicana, S.A., representing various American firearms and sporting goods firms, as he said, "selling all the ingredients necessary to a Latin American government."[11] This allowed him to travel throughout the region on deals from as far afield as Guatemala and Argentina. Applegate continued to follow close-combat needs during the Korean War in the 1950s. According to *Combat Forces Journal* reviewing Applegate's earlier book, although armed with many fine weapons, soldiers in Korea found that "combat often does develop in such close quarters and with such violence that the soldier must be prepared to fight with his fists or his bayonet, knife or anything else that is handy."[12] This was met in part by serving veterans with World War II experience and the enduring availability of the authoritative *Kill or Get Killed*.

Applegate in Mexico, where he participated in arms and security sales for more than a decade and a half. He developed close ties with both the American Embassy and Mexican officials.

An item demonstrated in Mexico was the ballistic shield to protect law enforcement personnel from armed opponents. Applegate was able to field test many items that were subsequently accepted as industry standards.

There was renewed demand for a fighting knife, met by sales of "British Commando" knives of the Fairbairn-Sykes design by Pasadena Firearms Company, which included an autographed picture of Fairbairn with each $3.95 purchase. Applegate followed up on his own ideas for improved knife design with help from Al Buck. However, before their plans were finalized the conflict ended, and more would come of this later. He did offer Randall-made knives to the South American market and continues his relation with their product line to this day.[13]

Applegate worked with Armamex, S.A., in the local assembly of U.S.-made parts into shotguns and small-caliber rifles in the mid-1950s to 1960s.[14] He represented various gun-related firms and handled consignments, including that for the Smith & Wesson .38 "Mexican Model" target revolver.

One of Applegate's experiences in Mexico was based on a preference at the time for carrying the Smith & Wesson .38 Safety Hammerless revolver for self-defense. This favorite was challenged when five shots failed to stop a machete-wielding assailant, who was brought down with the help of Applegate's companion firing a Colt M1911A1. Applegate discussed this with firearms expert W.H.B. Smith and together they lobbied Smith & Wesson company president Carl Hellstrom and designer Joe Norman for the use of the .38 Special in the 1952 Smith & Wesson Centennial revolver. This started Applegate's long association with Smith & Wesson projects and products.[15]

In another example of technical innovation, Applegate worked with Jack Canon in the development of what became the Glaser Safety Slug by getting it field tested in Mexico and introduced to law enforcement. The subsequent use of high-velocity, prefragmented pistol projectiles is seen in the current MagSafe line of products.[16]

Applegate felt this was an exciting period to be in Mexico, not that the United States defense attaché's office on the cocktail circuit would have noticed it. The Mexican military was polite but not open to "gringo" advice or aid. The head of the national arsenal was the man to know, as he could receive, sell, or ship arms and ammunition. "This was an active time of exile politics with revolutions in Guatemala and Cuba," he reminisced.[17] In his business, Applegate had some interesting problems to deal with, such as shipping ammunition to both sides of one conflict. Sales brought about further contacts with the Mexican military and police with the importation of riot control material and chemical agents.

Brevet Brigadier Applegate welcomes Mexican Secretary of Defense Gen. Augustin Olechea, Chief of Federal Security Police Manuel Rangel, and other dignitaries to a demonstration of Lake Erie riot control agents.

In 1961, Applegate's involvement expanded to include training Mexican police and army military police in riot control techniques, with a rank of brevet brigadier from the secretary of defense.[18] By then, the situation in Mexico City was critical because of communist-inspired riots that threatened to topple the government. This was particularly crucial after a student riot in Chilpancingo resulted in 60 or more dead and hundreds wounded when the army resorted to gunfire to disperse them. At that point Applegate was called in to assist Secretary of Defense Gen. Augustin Olechea to try to control disturbances without resorting to bloodshed. Applegate was employed to train Col. Miguel Torres Contreras' Military Police Regiment and Comdt. Frias Ramirez's Mexico City Police riot battalion. These developments were followed closely by the American Embassy's defense attaché's office.

With due regard for the complexities of Mexican factional rivalries and sensitivity to American influence, Applegate started by training one company-size force in tactics and techniques, and they in turn would train the other units in their respective organizations. The "Applegate Method" made extensive use of Lake Erie Chemical Company riot agents and the long riot baton. This was tried and tested in Mexico with satisfactory results and in turn recommended to the U.S. Army Military Police School.[19] Applegate recalled the concept of firepower alone not being the way to handle riots against civilians, and that the source of the firepower must be carefully controlled by responsible officers "in the rear of the baton contact line, where the armed men under them are able and disciplined to shoot on orders only. Same idea with grenade throwers and riot grenade and projectile launchers."[20]

Applegate believed the term "riot control" was a misnomer, for mobs and violent riots are first contained and then dispersed. Irrespective of terminology, riot control was a distinct and separate police science that demanded specialized, aggressive, and definite police action without undue political direction or hindrance. He stated it could no

Military policemen practicing riot control formations using the long baton advocated by Applegate. Communications, chemicals, and firearms are in the rear under direct control of the officer in charge.

longer be approached "in an emotional haphazard manner, involving only a few hours of 'rookie training' and a few departmental 'crash' refresher courses, as it oftentimes was. In the final analysis, the ultimate means for the maintenance of public order and stability is by use of force. Let it, however, be restrained, minimum necessary civil police force, operating under local authority."[21]

Applegate concluded that "destructive riots involving the use of all necessary police force, including firepower, can erupt at sporting events as well as political protests. Regardless of the reason for the crowd, police tactics to keep it orderly and law-abiding are always under the public spotlight. Poor police performance can result in injuries, bad publicity, departmental censure, leadership changes, and taxpayer expense."[22]

By 1963, the situation in the United States brought Applegate home to offer his services to jurisdictions ranging from Oxford, Mississippi, to Chicago, Illinois, during the turbulent 1960s. He was consulted by the provost marshal general of the U.S. Army and various police departments of state and local communities as "one of the nation's leading authorities on the fine and subtle art of riot control."[23] This experience found its way into an expanded edition of *Kill or Get Killed* to include the last word on mob control. By 1969, this led to a new book, *Riot Control: Materiel and Techniques*, that met the demands of security and police forces for information on what had become an all too common occurrence.

Applegate promoted the use of the long baton, the "batter's" helmet with face shield, and the bomb basket. Expertise with riot control agents saw discussions that resulted in the introduction of chemical mace and in the use of pepper sprays as alternatives to indiscriminate use of military CS and

A military police section advancing against a mob in the wedge formation. Gas shells have been launched, masks put on, and offensive action begun.

CN chemicals. In recent years, Applegate's testimony on law enforcement use of chemicals and other nonlethal weapons was sought by the U.S. Congress during the Bureau of Alcohol, Tobacco, and Firearms/Federal Bureau of Investigation standoff at Waco, Texas. Applegate's views on these events are guarded, as he puts it: "I have not and will not testify against law enforcement, even though I have problems with them."[24]

Based on his long-running experience in civil disorder, Applegate speculated that if the deterioration of urban areas continues because of crime, gangs, racial conflict, social and economic conditions combined with civil disturbance and acts of terrorism, permanent standby control forces would be required: "These will either be military or, preferably, civil law enforcement units organized, trained, and appropriately armed along military lines. It's an either-or situation." Due to changing political and social conditions, "the role of our military will probably now take on new dimensions in maintaining long-range domestic stability" and local police agencies are likely to find themselves in situations where it will probably be necessary to employ "paramilitary tactics and light weaponry"[25] to maintain law and order.

Applegate found time during this period to visit Korea and Vietnam, "but not as a combatant." Contributions proceeded with publications, new designs for fighting knives with Barry Wood, and renewed attention to combat shooting by law enforcement and the armed forces. By the mid-1970s, there was fresh demand for close combat as taught by Rex Applegate, met with a fifth edition of *Kill or Get Killed*, and Cassidy's research and writing about World War II knife- and gunfighting. Applegate also encouraged reprints of the classic *Get Tough* and *Shooting to Live* as aids to further study.[26]

Officials of the Batallon de Policia Militar along with Applegate and the special display of various tear gas grenades and equipment used in demonstrations.

This coincided with expanding public attention and notice of what was to be called the "modern technique of the pistol" that provided an *aimed-fire only* counterpoint to point shooting. Commenting upon the subsequent impact of the American Pistol Institute and International Practical Shooting Confederation (IPSC), Applegate concluded that acceptance of the two-handed stance with sighted shots for all armed showdowns "began with the formation of IPSC, and the Weaver experience, decades ago. The actual records of police firearms confrontations indicate that this system is largely responsible for the current police inability to perform adequately, at close ranges, under poor light conditions in lethal confrontations with armed criminals."[27] Applegate also challenged the assumption that the average policeman or soldier needed to be an enthusiastic sportsman before they could be any good in close combat with firearms: "The vast majority of military and police personnel are simply not interested in handgun shooting for recreational purposes. The type of personnel involved consider the handgun just another tool, infrequently employed."[28]

Another successful project saw Section Eight's wartime *Specialized Training in Foot Reconnaissance* released as *Scouting and Patrolling* in 1980. The publisher felt the topic was presented in a manner that was just as useful as it was when it was put together at Camp Ritchie many years before. The first printing was rapidly bought up by the U.S. Army Infantry and Ranger Schools and was personally used by the coauthor in training Marine division reconnaissance teams.

Not content to retire to fishing the Umpqua River in his native Oregon, Applegate drew on his experience with Fairbairn and knife design to eventually produce an innovative fighting knife in

Applegate with the legendary Bo Randall in the master cutler's shop in Orlando, Florida, at a meeting of two American blade pioneers with Paladin Press publisher Peder Lund. This relationship began in World War II with ideas for better fighting knives. (Lund collection.)

1980—this motivated by seeing publications about knives and knife fighting that were unrealistic—and concluded: "Most authors write about knives that are unsuitable for fighting, so they develop a technique around the weapon. This of course is next to useless."[29] This led him to reassert the "scientific" lessons already learned and published on the subject to accompany the availability of an effective fighting knife.

The resulting design was termed the Applegate-Fairbairn, after its principle antecedents. The blade is 6 inches long, 1 inch across at its widest, 3/16 of an inch thick at the spine, and has an overall length of around 11 inches. The blade's point, ricasso, and tang are massive compared to the original Fairbairn-Sykes knife (to prevent breakage, take a good edge, and to both cut and slash). The forward curved crossguard is of 1/4 inch brass to catch a parry and to fit the thumb. Plastic was used for the oval-shaped handle, with longitudinal grooves and thumb flats. Balance is in the handle to keep the blade point up in the hand. Various sheaths reflect the use of new materials for long life and hard wear.[30]

The demand for the Applegate-Fairbairn knife was met by various craftsmen (T.J. Yancey, Al Mar, Blackjack, Bill Harsey, Butch Vallotton) and is in current production by Boker Baumwerk of Solingen, Germany. Other ideas ensued, including prototypes of the Applegate combat smatchet and mini-smatchet in 1988. The Applegate combat folder was produced by Gerber Legendary Blades of Portland, Oregon, in 1996, and a "covert" folder followed in 1998. Years of interest in these useful edged weapons earned Applegate a well-deserved place in *Blade* magazine's Cutlery Hall of Fame.[31]

Shooting interests continued as well with Applegate's "Guns of Famous Shooters" collection and documentation of the exhibition shooters he knew as a boy. This showcased firearms and

Rex Applegate with the publications and hardware that reflect unique contributions and expertise up through the present day, based upon the wartime work and experience documented in this book.

accessories owned by Peret, the Topperweins, FitzGerald, McGivern, Askins, Jordan, Wesson, and even a Colt ordered by Sykes for the Shanghai Municipal Police.[32] One result was the location and presentation of a number of videos of rare film footage of these shooters at work.

Other opportunities followed from surprising sources that once more brought Applegate's experience into use. In 1984, Marine Gen. Wallace M. Greene recalled the use of Shanghai-developed point shooting before and during World War II. This enduring focus on practical gunfighting saw works by Fairbairn, Sykes, and Applegate published by the Marine Corps as FMFRP12-81 *Shooting to Live* (1990) and 12-80 *Kill or Get Killed* (1991), which spotlighted their approach to close combat. The director of the Marine Corps Combat Development Command's Warfighting Center, Maj. Gen. Matthew P. Caulfield, felt the "detail, techniques, and training procedures presented will enhance small unit training, and every unit involved in the above activities should have copies."[33]

More roundabout influence was witnessed in the U.S. Army FM21-150 *Combatives* (1992). A direct claim was made that asserted that the Marine Corps' Linear In-Fighting Neural-Override Engagement Program in FMFM 0-7 *Close Combat* (1993) was based on the previous efforts since World War II. One senior military instructor stated that it was a derivative fighting style based on the earlier "techniques of Fairbairn, Applegate, Sykes, Styers, Biddle, and Echanis."[34] The basis for this assertion needed context and comment, and this volume preserves the specifics of the techniques that the continuity claimed is made more actual than apparent.

In 1993, Applegate responded to general dissatisfaction with the path that "practical" shooting was taking, entering the debate over aimed versus point shooting by calling upon his own background and making it available to a new generation of American law enforcement and military shooters. In October 1994, he published a provocative article in *Law and Order* that stated "bullseyes and silhouettes don't shoot back" and that law officers should be given handgun training with and without the use of sights.[35] He followed this up by locating and making a video called *Point Shooting* from the earlier U.S. Army training film (FB152) and produced the updated *Shooting for Keeps* video.

Created were new forums to teach an old method and "the results in better field performance of officers has been most gratifying" to Applegate.[36] This was as Bruce K. Siddle (PPCT Management Systems), Bill Burroughs (Smith & Wesson, SIG Arms), and Ken Hackathorn and Bill Wilson (Wilson Combat) saw point shooting's self-evident usefulness and promoted it in their own curriculums of firearms and self-defense instruction.

Intrigued by these ongoing events, Steve Barron and Clyde Beasley of the Department of Public Safety, Hocking College, Ohio, made point shooting the focus of their resident peace officer basic training in 1995. Barron had concluded that if gross motor movement techniques were employed in self-defense measures to be used under stress, then why were they trying to teach fine motor skills for using firearms under similar situations. The thrust of this program noted that point shooting was to entirely replace previous instruction in only the Weaver stance for aimed shooting with the "isosceles stance combined with a convulsive grip and a push-pull movement involving holding the gun-holding hand and support hand for firming up the sight picture at the time of firing."[37]

Applegate called upon the research and experience documented here to advocate that "point shooting covers the most practical, easily learned, and retainable technique for shooting the handgun, in close-quarter, life-threatening situations." Point shooting, as Applegate urged it, involves a stiff arm, locked elbow and wrist, plus a convulsive grip on the weapon. The weapon is raised to eye level like pointing a finger and instantaneously fired. Eyes always on the target, not on the sights. The gun is raised with a rigid arm like a pump handle and the arm pivots at the shoulder. The system is accurate at distances up to 15 yards. It is faster and works in all types of light situations.[38] The technique can be summarized as: 1) facing the threat with both eyes open; 2) raising the weapon to eye level and instantly firing with no front sight picture; 3) having a convulsive grip on the handgun; and 4) mostly firing and training from a forward, aggressive crouch. The two-handed isosceles stance is also recommended for all types of sighted fire when distance, light, cover, and time permit. Supporting elements include the use of suitable firearms (large-bore double-action automatics), conditioned firing responses, both eyes on the point-of-aim for center of mass hits, multiple shots, personnel targets, and realistic scenarios. Applegate concludes that "these techniques have been, and are, historically validated and documented."[39]

Some observations can be made in regard to the discussion of deliberate versus point shooting for close combat. Differences seem to depend upon poorly agreed upon terms and concepts, inadequate training, and a lack of useful integration that recognizes that each approach works under different circumstances. Rex Applegate brought one to its greatest refinement, Jeff Cooper and others have championed the other; both meet at the shifting point of transition on a continuum defined by the terrain and situation.

As this was going on, Applegate successfully ran for the National Rifle Association's board of directors and received the 1996 Outstanding American Handgunner Award for his lifetime contribution. Applegate continues to provide the "real thing" to those Americans who find themselves in harm's way.

History is selective, in this case to create an account of close-combat techniques taught at a certain point in the past that could be of use at present and in the future. This provides documented theory and practice to fill the gap in training and doctrine that this field cyclically falls into. While not a replacement for known-distance or martial arts competition and qualifications (which may be irrelevant), it is a necessary and logical extension to cover an area often ignored in individual training or subject to misplaced specialization.[40]

"It all depends on the terrain and situation," instructors used to say, and here are classics of the field of close-quarter battle. If you are in the military, law enforcement, or intelligence work, then you should be cognizant of this information, which will give you an edge when far away from instructors, formal schools, and facilities (as most are).

It is critical to keep in mind that soldiers and policemen have different roles and responsibilities. One serves and protects citizens while the other closes with and destroys the enemy; don't mix this up. Applegate noted early on that the soldier "must be trained and indoctrinated in the offensive. Combat between armies is only won by offensive tactics. The law enforcement officer has a different problem. He must first master restraint and manhandling tactics. He must also be able, under extreme or necessary circumstances, to take strong defensive or offensive action."[41]

Taken as a whole, over a half century since World War II's end, Applegate has come full circle back to the wartime experience presented here. He wrote what is a fitting conclusion:

> I have purposely avoided laying out specific, detailed training programs. Each organization—military or civilian—has its own problems, some phases of training demanding more emphasis than others. Although . . . pointed toward the training of large groups of men, I hope that those individuals who have sufficient interest to study it will, as a result, find themselves better prepared should they suddenly find themselves opposed by a killer.[42]

COLONEL APPLEGATE'S PARTING SHOTS

To pick up the story begun in the introduction, after the war I lived abroad and became involved in other matters such as riot control, and I had very little to do in respect to handgun training. However, over the past few years I began to notice some disturbing trends. My concerns were substantiated three years ago when I attended a lecture at a law enforcement conference on advanced handgun shooting.

The young instructor said you put your foot forward here, and put your other foot back here, bend your elbow here, put your other hand here, control your trigger finger this way, always shoot with two hands, and use the sights for all types of close-quarters shooting. He certainly was not doing much thinking about the kind of shooting the cop encounters almost every day on the street and in dark alleys. Anyway, at the end of the class a SWAT team member of the audience got up and asked, "Young man, why do you think this is the best way to shoot at close quarters in a gunfight, or for training a policeman to shoot in the street?" The instructor said, "Because 50 of the world's greatest combat competition shooters shoot this way."

That's when I really took notice. It was a real shock to me that the Weaver style of two-handed, sighted shooting and training had been going on for so long, almost entirely to the exclusion of other methods, and that there was almost no police or military close-quarter training being given that did not involve the use of the sights.

Currently, the combat competition type of instruction is being given all over the United States, mostly by civil law enforcement instructors, the feds, and the military. Reports show that most trainees can't hit their ass under stressful, close-quarter firefight situations.

The best records we have are in New York Police Department (NYPD) statistics. Every year for the past 10 years, it has put together statistics on every firefight that its officers have engaged in with perpetrators, as it calls them.

The statistics are very complete as to the details of these firefights. The time of day or night; whether the officer was kneeling, sitting, standing, aiming, using two hands; what the light conditions were; what the ranges were; the number of shots fired by the perpetrators—the whole thing. These records also summarize each firefight by a brief description of the event. You couldn't ask for better data to work from.

This yearly data covering 30,000 to 40,000 policemen also generally applies to the rest of the United States as well. Every year the NYPD's average of hits in firefights is somewhere between 14 and 16 percent. Now that's pretty sad. I'm also talking about police elsewhere in the nation who are on the streets using the same techniques.

Now maybe I'm wrong or being illogical. I know there are a lot of problems about what a given police administration directs and that many police departments get locked into a handgun training program that only the chief wants or the politicians require or limited finances permit. But you would think that, even in New York City, over a 10-year period when there was a system of training that was failing, after the first two or three years the people responsible would say, "Gosh, my guys are shooting only 14 to 16 percent according to our own records. We must be doing something wrong." Well, nothing has happened, and it goes on and on.

Statistics indicate that 60 to 80 percent of all police handgun firefights with criminals take place at distances of 20 feet or less, under conditions in which there is little or no light to see the sights, no time to see the sights, or no opportunity to use the sights. Irrespective of this fact, what's happening is that almost all civil and federal law enforcement agencies, and our military, do not train in the point-shooting technique. Accordingly, too many trainees miss at close range due to the lack of, or improper, training.

Everybody knows that you can shoot a handgun better with two hands—if you can use the sights and have time to aim—than you can by using one hand. But most of the time this does not apply to the cop or soldier who has his adrenaline pumping and is in a dark alley or on a battlefield facing his own or his enemy's blinding muzzle flash along with a loud "bang" under a combat stress, life-threatening situation.

In this time of political correctness, if I were an instructor and I had a class of recruits of both sexes who were a mixture of different ethnic backgrounds and who were from urban areas and had no firearms background, I would start out the handgun course with required familiarization and safety procedures. Taking into account student interest and personal survival instincts, I would start out with close-combat point shooting first. I would teach two-handed, aimed shooting with the isosceles stance only—equal time to both techniques.

Regardless of how long you train in the Weaver stance method, statistics show that most shooters revert to isosceles under life-threatening situations. Training with the isosceles, I would advocate a convulsive grip and a slight push-pull movement to firm up the sight picture just before firing.

Point shooting is simple. It's basically a convulsive grip on the firearm, a rigid wrist, and a rigid elbow and shoulder pivot point. You raise your gun hand to eye level and go "bang"—no separate trigger finger action, no recoil problems. It's like pointing your finger. You can teach almost anyone to shoot this way with what I call "head hitting" accuracy.

You may have wondered about why there are so many police-criminal firefights in which a great number of police shots are fired at ranges of less than 20 feet resulting in no hits or maybe someone getting hit in the big toe. One major reason for this is that the Weaver stance-trained policeman was probably using one hand and under combat pressure had a convulsive grip on the weapon and shoved his weapon at the target.

When this happens, either from a gun-in-hand raised-pistol position or in conjunction with a draw, the barrel of the handgun points down when the gun arm reaches its full-thrust firing position. This is due to the combination of the convulsive grip and gun design as they relate to the angle of grip to the barrel. "Thou shalt not shove thy handgun"—lift it to eye level in the same manner as you would raise and point your finger.

At this time an extensive program on point shooting is taking place at Hocking College in Nelsonville, Ohio. Initial reports indicate that it is an outstanding success. The police science division of the college is training new police recruits and students in this discipline and retraining officers who have been previously trained in Weaver only. All are being required to meet the Ohio State Police standards. Each type of student is being compared against the other. Statistics, videos, and other related data are available. I might also add that for training emphasis they are using some handguns on which the front sight has been filed off. Serious law enforcement trainers can contact Steve Barron of Hocking College for the most recent hands-on training experience report on this vital handgun combat technique.

Several years ago, after it had disappeared, I tried to locate film FB152, *Combat Use of the Handgun*. Six months later I finally found a copy at Norton Air Force Base, in California. After my discovery, the army had it declassified and sent me a copy. I took it to Paladin Press, and now it's in the form of a video called *Point Shooting*. I did an introduction and background covering, more or less, the things I have said here.

Police trainees at the close position at Hocking College in 1997. They are practicing Applegate's close-combat shooting techniques. (Photo courtesy of Steve Barron.)

Hocking College's range in operation. Point shooting's application to current police and military situations can be met with very little change in anything except attitude and recognition that "bull's-eyes don't shoot back." (Photo courtesy of Steve Barron.)

EDITOR'S NOTE

As we go to press with this, the latest of Rex Applegate's works, we have received here in Paladin's editorial offices a letter from Chief D. Watkins of the California Highway Patrol's Border Division stating that "The California Highway Patrol is currently utilizing point shooting as its close-quarters combat shooting method. This method of close-quarters shooting is based upon using gross motor skills, which is consistent with all the research that has been done in this area. Our officers are readily adapting to point shooting. We see immediate increases in both speed and accuracy levels. This system of close-quarters combat shooting will undoubtedly increase officer survival and save lives." Chief Watkins went on to say that point shooting is a "relevant way to deal with the spontaneous, close-quarters gunfight situations that our officers face in the course of their duties."

This letter highlights an important event in the history of point shooting and clearly demonstrates that the old ways of the OSS have stood the test of time.

—B.N.

This year I was featured in a Paladin Press video called *Shooting for Keeps*. It covers in detail point-shooting techniques and brings them up to date.

In conclusion, although I'm 83 years old and my thoughts may not be up to date, everything that I see going on today has not altered my views at all on the value of unsighted point shooting for close-quarters handgun combat.

Proper training in combat point shooting achieves quicker expertise, does not necessitate so much retraining to maintain proficiency, saves more police lives, and takes more criminals permanently off the streets. There has been too much concentration on what is called the "new modern pistol technique" instead of on what has been historically proven in combat.

PUBLISHER'S NOTE

Col. Rex Applegate died on July 14, 1998, as we were sending this book to press. Colonel Applegate was a mentor, colleague, and friend to many members of the firearm, knife, and law enforcement communities. The Colonel leaves behind a body of work that has instructed and influenced generations of fighting men and women around the globe. He will be missed.

Col. Rex Applegate, 1914-1998.

ENDNOTES

1. While an instrument of the Joint Chiefs of Staff, the OSS was excluded from South America and areas of the Pacific theater. The OSS was fixed to operations on the mainland of Southeast Asia, while the Southwest Pacific Area (SWPA) came under Gen. Douglas MacArthur's General Headquarters Allied Intelligence Bureau (AIB). Observed assistant chief of staff G2 Maj. Gen. Charles A. Willoughby, the AIB was the counterpart of the OSS with SOE representation, and despite the postwar trends to belittle the old-time military and naval intelligence services, the "AIB functioned as a military service on a budget that could not compare with the enormous financial resources of the OSS. Nevertheless, it operated successfully behind enemy lines." (*Intelligence in War: A Brief History of MacArthur's Intelligence Service, 1941-1951* [Tokyo: USAFPAC, n.d.], 8.)
2. Hyde, *Room 3603*, 229.
3. Applegate, Introduction, *Commando Dagger*, viii.
4. Cassidy, *Quick or Dead*, 106.
5. William E. Fairbairn, *Get Tough!* (Boulder, CO: Paladin Press, 1979), vi.
6. Dennis Martin, "The Marcus Wynne Interview," *Fighting Arts International*, December 1995, 31-32.
7. A topic for further study would be the underlying theoretical research found in the works of Louis A. La Garde, Julian S. Hatcher, S.L.A. Marshall, David Ben-Asher, Bruce K. Siddle, Dave Grossman, and Gregory B. Morrison.
8. Rex Applegate, personal communication, 30 October 1995.
9. For example, see FB152 *Combat Use of the Handgun*; FM23-35 *Pistols and Revolvers* (Washington, DC, USGPO, 1953); and A.M. Kamp, Jr., "Pistol Training for Combat," *American Rifleman*, September 1954, 17-19, 79. This same course of fire was in use by the U.S. Navy through the 1970s.
10. Rex Applegate, "Training Combat Pistolmen," *American Rifleman*, July 1944, 4.
11. Joe McGinniss, "The Passing Scene: Pro on Keeping Peasants in Line," *The Philadelphia Inquirer*, 1 April 1968, 23.
12. "Hand-to-Hand," *Combat Forces Journal*, October 1951, n.p.
13. Chuck Karwan, "Born Again Combat Legend," *Tactical Knives*, January 1998, 21; Gaddis, *Randall Made*, 160-162.
14. Dickey, "Connoisseur," 41.
15. Dickey, "Connoisseur," 41; Clapp, "American Hero," 72; Chuck Karwan, *Col. Applegate and His Knives* (Portland, OR: Gerber, 1997), 15-16.
16. Karwan, *Applegate and His Knives*, 8.
17. Rex Applegate, personal communication, 11 September 1997.
18. Rex Applegate, *Mexican Riot Control*, September 1992. Applegate still maintains insights and interest in politics south of the border; for example, Rex Applegate, "Time Bomb: The Mexican Threat to US Security," *Informed Source Intelligence Newsletter*, March-April 1997, 10-13.
19. Rex Applegate, "The Organization and Tactics of Professional Riot Control Forces," *Army*, March 1963, 53-61.
20. Applegate, *Mexican Riot Control*, 3.
21. Rex Applegate, *Riot Control: Material and Techniques* (Boulder, CO: Paladin Press, 1981), 11.
22. Rex Applegate, Foreword, *Police Crowd Control*, by Charles Beene (Boulder, CO: Paladin Press, 1992), vii.
23. McGinniss, "The Passing Scene," 23.
24. James Bovard, "Convoluted Trail of Waco Explanation," *The Washington Post*, 19 April 1995, A15; Karwan, *Applegate and His Knives*, 8; Rex Applegate, personal communication, 11 September 1997.
25. Rex Applegate, Foreword, *An Infantryman's Guide to Combat in Built Up Areas*, by (Boulder, CO: Paladin Press, 1994), ix.
26. Paladin Press used Applegate's copy of the Fairbairn and Sykes' book for reproduction.
27. Rex Applegate, personal communication, 7 August 1996.
28. Rex Applegate, personal communication, 7 August 1996, in response to Jeff Cooper, "Jeff Cooper on 20 Years of IPSC," *American Handgunner*, September/October 1996, 92-93, 108-111; Rex Applegate, personal communication, 15 July 1997.
29. Virginia Thomas, "Applegate's Knife Combat," Paladin Press, 13(4) 1988, 17.
30. Specifications vary depending upon manufacturer.
31. Karwan, *Applegate and His Knives*, 7-20; Leroy Thompson, *Commando Dagger* (Boulder, CO: Paladin Press, 1985), 157-164; Thomas, "Applegate's Knife Combat," 16-17; Karwan, "Born-Again Combat Legend," 18-21, 60.
32. Clapp, "Colonel Rex Applegate," 73.
33. FMFRP12-80, *Kill or Get Killed* (Washington, DC: HQMC, 1991), i.
34. Ronald S. Donvito, "Close Combat: The LINE System," *Marine Corps Gazette*, February 1995, 24; as early as 1966, the Marines had adopted a program designed by Fairbairn SMP colleague Dermot M. O'Neill.
35. Rex Applegate, "Combat Firing with Handguns," 8 August 1994; "Bullseyes and Silhouettes Don't Shoot Back," *Law*

and Order, October 1994, 1-4; "Combat Handgun Training," 12 December 1994; "Basic Instinct," *Soldier of Fortune*, March 1995, 48-51, 67.

36. Karwan, *Applegate and His Knives*, 14.
37. Rex Applegate, personal communication, 30 September 1996; Steve Barron, personal communication, 1, 3 July 1997; Hocking College Spring 1997 courses, 24-25; Bill Clede, "Point Shooting," *Law and Order*, October 1996, 128-131; Steve Barron, "Point Shooting Is Alive and Well," *Law and Order*, September 1997, 40, 42-44.
38. Rex Applegate, personal communication, 6 March 1997.
39. Rex Applegate, personal communication, 15 July 1997.
40. Gregory B. Morrison, "Validation: Police Handgun Instruction's Unmet Challenge," *The ASLET Journal*, May/June 1996, 11-15; "The FBI and Police Firearms Training," 15 March 1996; these provide background to law enforcement shooting as it developed in the United States through World War II based upon material presented in his dissertation on the subject. The impact of military wartime experience remains to be explored.
41. Applegate, *Kill or Get Killed*, xi.
42. Applegate, *Kill or Get Killed*, xii.

BIBLIOGRAPHY

Applegate, Rex. "Unarmed Offense," *Infantry Journal*. March 1943, 22-27.

—"Unarmed Offense," *Infantry Journal*. April 1943, 46-52.

—"Handgun Offense," *Infantry Journal*. August 1943, 53-54.

—"Knife Fighting," *Infantry Journal*. December 1943, 46-50.

—*Kill or Get Killed*. 1st ed. Harrisburg, PA: Military Service Publishing Company, 1943.

—"Training Combat Pistolmen," *The American Rifleman*. July 1944, 14-16.

—"House of Horrors," *The American Rifleman*. August 1945, 20-23.

—"The Organization and Tactics of Professional Riot Control Forces," *Army*. March 1963, 53-61.

—"The Long Riot Baton," *Law and Order*. March 1964, reprint.

—*Crowd and Riot Control*. 1st ed. Harrisburg, PA: Stackpole, 1969.

—*Kill or Get Killed*. 5th ed. Boulder, CO: Paladin Press, 1976.

—*Scouting and Patrolling*. Boulder, CO: Paladin Press, 1980.

—*Riot Control: Materiel and Techniques*. Boulder, CO: Paladin Press, 1981.

—*Fast and Fancy Shooters*. Scottsburg, OR: Wells Creek Knife & Gun Works, 1989, videocassette.

—*The Fabulous Topperweins*. Scottsburg, OR: Wells Creek Knife & Gun Works, 1992, videocassette.

—*Mexican Riot Control*. Scottsburg, OR: Author, September 1992.

—*Combat Use of the Double-Edged Fighting Knife*. Boulder, CO: Paladin Press, 1993.

—"Bullseyes and Silhouettes Don't Shoot Back," *Law and Order*. October 1994, reprint.

—"Basic Instinct: Point Shooting Not Dead Yet," *Soldier of Fortune*. March 1995, 48-51, 67.

—*Point Shooting*. Boulder, CO: Paladin Press, 1995, videocassette.

—"Shooting for Keeps," *Tactics*, 8(2). 1996, 4-10.

—*Shooting for Keeps*. Boulder, CO: Paladin Press, 1996, videocassette.

—"Rex Applegate on Combat Shooting," *Handguns*. January 1997, 70-75, 85.

—"The Mexican Threat to U.S. Security," *Informed Source*. March-April 1997.

—"Bill Jordan, the Fast Man with the Slow Drawl," *Handguns*. May 1998, 50-54.

Applegate, Rex, & Walter D. Randall, Jr. *The Fighting Knife*. Orlando, FL: Author, 1969.

Applegate, Rex, Tom Campbell, Wiley Clapp, & Chuck Karwan. *Manstoppers*. Boulder, CO: Paladin Press, 1991, videocassette.

Applegate, Rex, Kelly Yeaton, & Samuel S. Yeaton.

The First Commando Knives. Williamstown, NJ: Phillips Publications, 1996.

Applegate, Rex, & Michael D. Janich. *Bullseyes Don't Shoot Back*. Boulder, CO: Paladin Press, 1998.

Askins, Charles, Jr. *The Art of Handgun Shooting*. New York: A.P. Barnes & Co., 1939.

Baum, G.M. "Common Sense Training with the Service Pistol," *U.S. Naval Institute Proceedings*. August 1923, 1309-1315.

Cassidy, William L. *The Complete Book of Knife Fighting*. Boulder, CO: Paladin Press, 1975.

Cassidy, William L. *Quick or Dead*. Boulder, CO: Paladin Press, 1978.

Cunningham, Eugene. *Triggernometry: A Gallery of Gunfighters*. Caldwell, ID: Caxton Printers, 1934.

Fairbairn, William E. *Instructions and Conditions of Practice for the .45 Colt Automatic Pistol*. Shanghai: Shanghai Municipal Police, [1925].

—*Manual for the Thompson Sub-Machine Gun*. Shanghai: Shanghai Municipal Police, 1940 (GHCA).

—*All-In Fighting*. London: Faber & Faber Ltd., 1942.

—*Hands Off!* New York: Appleton-Century Co., 1942.

—*Get Tough!* New York: Appleton-Century Co., 1943.

—"Shock and Riot Police." Washington, DC: OSS, June 1943 (NARA).

—"Gutter Fighting: the Fighting Knife." Washington, DC: OSS, 11 February 1944 (NARA).

Fairbairn, William E. and Eric A. Sykes. *Shooting to Live with the One Hand Gun*. Edinburgh: Oliver & Boyd, 1942.

Fairbairn, William E. "Battle Firing for Those Who Want to Live." n.d. (GHCA).

—"Close Combat." n.d. (GHCA).

—"Close Combat." Washington, DC: OSS, n.d. (NARA).

FitzGerald, J. Henry. *Shooting*. Hartford, CT: G.F. Book Co., 1930.

Greene, Wallace M., Jr. "The Employment of the Marine Rifle Company in Street Riot Operations," *Marine Corps Gazette*. March 1940, 49-62.

—"Shanghai 1937," *Marine Corps Gazette*. November 1965, 62-63.

—"The Quick or the Dead," *Marine Corps Gazette*. August 1984, 67-72.

Grossman, Dave. *On Killing*. New York: Little, Brown & Company, 1995.

Hatcher, Julian S. *Textbook of Pistols and Revolvers*. Marines, NC: Small Arms Technical Publishing Co., 1935.

Jordan, William H. *No Second Place Winner*. Concord, NH: Police Bookshelf, 1965.

La Garde, Louis A. *Gunshot Injuries*. New York: William Wood and Company, 1916.

Marshall, S.L.A. *Men Against Fire*. Gloucester, MA: Peter Smith, 1978.

McGivern, Ed. *Fast and Fancy Revolver Shooting*. Lewistown, MT: Author, 1938.

Morrison, Gregory Boyce. "A Critical History and Evaluation of American Police Firearms Training to 1945" (dissertation). Irvine, CA: University of California, 1995.

—*The Modern Technique of the Pistol*. Paulden, AZ: Gunsite Press, 1991.

Office of Strategic Services. *OSS Training Group* (film). Washington, DC: Field Photographic Unit, 1942 (NARA).

Office of Strategic Services. *OSS Weapons*. Washington, DC: Research and Development Branch, June 1944 (NARA).

O'Neill, Dermot M. "Hand-to-Hand Combat." Quantico, VA: Marine Corps Schools, November 1966 (MCCDC).

Siddle, Bruce K. *Sharpening the Warrior's Edge*. Millstadt, IL: PPCT Research Publications, 1995.

—*Scientific and Test Date Validating the Isoceles and Single-Hand Point Shooting Technique*. Millstadt, IL: PPCT Research Publications, 1996.

Stavers, Stephen. "Snap Shooting in Close Combat," *Infantry Journal*. December 1944, 30.

Sykes, Eric A. "Shooting Comments." n.d. (GHCA).

Yeaton, Samuel S. "Modern Pistol Training for the Marine Corps," *Marine Corps Gazette*. August 1937, 36-38.